AF710198

in Raffaelle and in Perugino the religious spirit smiled through its tears (Hegel).

Metaphysics "We must renounce metaphysics if we would mould our lives to artistic perfection.
For Philosophy serves culture not by the fancied gift of absolute or transcendental knowledge but by suggesting questions which help us to detect the passion and strangeness, and dramatic contrasts of life."
Yet surely he who sees in colour no mere delightful quality of natural things but a spirit dwelling on things is in a way a metaphysician.

Oscar Wilde's Oxford Notebooks

To Alan love Oscar

Portrait of Oscar Wilde taken 19 March 1881.

OSCAR WILDE'S OXFORD NOTEBOOKS

A Portrait of Mind in the Making

* * *

Edited, with a commentary, by
Philip E. Smith II
and
Michael S. Helfand

New York Oxford
OXFORD UNIVERSITY PRESS
1989

Oxford University Press

Oxford New York Toronto
Delhi Bombay Calcutta Madras Karachi
Petaling Jaya Singapore Hong Kong Tokyo
Nairobi Dar es Salaam Cape Town
Melbourne Auckland

and associated companies in
Berlin Ibadan

Copyright © 1989 by Oxford University Press, Inc.

Published by Oxford University Press, Inc.
200 Madison Avenue, New York, New York 10016

Oxford is a registered trademark of Oxford University Press

All rights reserved. No part of this publication may be reproduced, stored in a retrieval system, or transmitted, in any form or by any means, electronic, mechanical, photocopying, recording, or otherwise, without prior permission of Oxford University Press.

Library of Congress Cataloging-in-Publication Data
Wilde, Oscar, 1854–1900. [Oxford notebooks]
Oscar Wilde's Oxford notebooks : a portrait of mind in the making / edited, with a commentary, by Philip E. Smith II and Michael S. Helfand.
p. cm. Includes index.
ISBN 0-19-505133-5
1. Wilde, Oscar, 1854–1900—Notebooks, sketchbooks, etc.
2. Wilde, Oscar, 1854–1900—Knowledge and learning.
I. Smith, Philip E. II. Helfand, Michael S.
III. Title. IV. Title: Oxford notebooks.
PR5812.S55 1989 828'.803—dc19 87-30292 CIP

The jacket and frontispiece illustrations of Oscar Wilde courtesy of the William Andrews Clark Memorial Library, University of California.

The section entitled "Anarchy and Culture: The Evolutionary Turn of Wilde's Cultural Criticism" originally appeared in *Texas Studies in Literature and Language,* Vol. 20, No. 2 (Summer 1978). It is reprinted in revised form by permission of the University of Texas Press.

9 8 7 6 5 4 3 2 1

Printed in the United States of America
on acid-free paper

For Mary Tucker Smith.
For the Helfands, living and unliving, who made the work possible and endured its creation.

Preface

The critical essay and the notebooks which follow it describe and present a far different Oscar Wilde than the dandy, aesthete, and homosexual who has become a myth for modernist sensibilities. The Wilde of the notebooks is a precocious Victorian humanist, an Oxford undergraduate studying in the *Literae Humaniores* program and later, a postgraduate competing for a faculty position at Oxford, immersed in a year's research and writing for the Chancellor's English Essay Prize of 1879. He studies a number of controversies which concerned contemporary intellectuals; for instance, evolution and human descent, historical criticism, and the opposition of philosophical idealism and materialism.

In our critical essay we reinterpret Wilde's work in the context of this moment in nineteenth-century intellectual history. We argue that the foundations of his later work were laid in the reading and writing he did during and shortly after his years at Oxford. Our primary evidence for this claim comes from Wilde's previously unpublished Commonplace Book and Notebook Kept at Oxford, which we have edited for this volume. These neglected documents contain the records of Wilde's education and serious reading in the late 1870s and early 1880s, as well as raw materials, drafts, and fragments used for his later writing. We have spent years identifying many of Wilde's sources for notebook entries; they include writers often associated with Wilde—Plato, Aristotle, Baudelaire, Swinburne, Pater, Ruskin—as well as humanists, philosophers, and social scientists not usually associated with him—H. T. Buckle, W. E. H. Lecky, Benjamin Jowett, William Wallace, T. H. Huxley, John Tyndall, W. K. Clifford, E. B. Tyler, Herbert Spencer, Ernest Renan, Theodor Mommsen, J. A. Symonds, G. W. F. Hegel, and many others.

We argue that Wilde's later critical and creative works have been misunderstood and undervalued because critics and scholars have not taken Wilde's education as seriously as he did.[1] Wilde's aestheticism, usually thought of as derived from Pater, Arnold, Ruskin, and the French decadent poets, is shown by the notebooks to be based on a carefully reasoned philosophical and political stance, a synthesis of Hegelian idealism and Spencerian evolutionary theory which fundamentally shaped his criticism and fiction. We believe that publication and interpretation of this material will provide the basis for a revaluation of Wilde's significance in the history of literature and criticism.

Our critical essay is divided into three sections: "The Text," "The Context of the Text," and "The Text as Context." In "The Text" we present a brief physical description of the manuscripts, historical evidence for dating, and the editorial principles which have guided our practice. We begin "The Context of the Text" by describing the vital intellectual influence of Wilde's parents and their circle in Dublin and of his later association with J. P. Mahaffy at Trinity College. Even before he entered Oxford, Wilde was introduced to subjects he continued to study: classical literature, philology, history, the newly developing social sciences, and the history of ancient and modern philosophy. While he may have abandoned some early theories, he nevertheless derived from them an expectation that any account of human experience had to have systematic and synthetic explanatory power. This theoretical expectation can help explain Wilde's interest in the theories of two of his teachers at Oxford, Max Müller and John Ruskin. In different ways both men attempted to accommodate their idiosyncratic idealist beliefs with certain findings of modern science about historical philology, myth, and the racial inheritance of morality, imagination, and the "art-gift."

At Oxford Wilde also found in Hegelian philosophy a variety of idealism better suited than Müller's and Ruskin's to incorporate the materialist assumptions and findings of science—especially evolutionary theory. Two of the Oxford Hegelians, William Wallace and Benjamin Jowett, recognized, taught, and wrote about the affinities between the theories. Wilde tested their suggested synthesis of idealism and science primarily by investigating the philosophical writings of critical materialists like Huxley, Tyndall, and Spencer. He found support for the synthesis in the scientists' own admission of the limitations of their knowledge and theories. However, he also found in Spencer's evolutionary theory—and especially in W. K. Clifford's revision of it—a scientific explanation for the development in history of Hegel's "Idea" namely, the inheritance of acquired characteristics. Wilde developed a position very close to Clifford's, but he added to the latter's incipient reconciliation of idealism and positivism the conception of absolute idea slowly evolving toward perfection in history. The inheritance of acquired characteristics provided the mechanism for this progressive development, and the assumption of a unity between mind and matter, ego and non-ego, "mind-stuff" and the organic molecules of the brain gave him the physical and spiritual location in individuals for absolute mind. These ideas also supported Wilde's belief in a racialist theory of cultural improvement. By adopting the Spencer-Clifford line of evolutionary theory, Wilde embraced an essentialist tradition of biological explanation (organic purposiveness) which began with Aristotle. His revisions to the tradition in his notebooks, criticism, and fiction demonstrate his critical practice of Hegelian theory.

Wilde's philosophical position led him to theoretical disagreements with Pater concerning philosophy and the function of art. In the notebooks Wilde copied from, and disagreed with, Pater's essay "Winckelmann." Other comments in the notebooks show that in the intellectual controversy regarding the interpretation of Aristotle's *Ethics,* Wilde adopted an idealist position, while Pater joined the philosophical materialists. Wilde therefore differed with Pater over the nature and power of art in society. He believed that Aristotle's *theoria* (ethically informed contemplation) yielded wisdom, the true basis for the good and spiritual life. Pater, however, believed that *energeia* (heightened consciousness) was Aristotle's highest good. For Wilde imaginative contemplation provided the basis for ethical, social behavior and was in itself the good, a manifestation in the individual of Hegel's idea and Aristotle's unmoved mover. For Pater contemplation was a self-centered act, informed by a materialist and utilitarian ethic in which the good is defined as "being present always at the focus where the greatest number of vital forces unite in their present energy."[2] Wilde differed with Pater, in short, over the fundamental theoretical issue of evolutionary idealism.

In "The Text as Context" we use Wilde's synthesis and the notebooks to reinterpret his published criticism and fiction. He relied on them to write "The Rise of Historical Criticism" (1879); this early essay, generally ignored by Wilde's critics,[3] is a Hegelian description of the origins and development of a scientific historical method. He argues that the idea of uniform sequence came to the Greeks by inspiration and was developed into a self-conscious critical method through the dialectical interplay of materialist and idealist philosophers and historians.

In his lectures in America, which also draw upon the notebooks, Wilde again quotes from, and implicitly disagrees with, Pater's position. Upon this disagreement he builds his recommendations for an aesthetic and ethical approach to integrating the arts with the activities of life and work. He adopts and modifies the Hegelian paradigm for the historical development of art to explain the English Arts and Crafts and aesthetic movements as rebirths of the Hellenic spirit.

Intentions, Wilde's collected major criticism, represents the fullest statement and development of his synthesis. We study the essays individually, locating statements and uses of Hegelian and evolutionary ideas. Our analysis of his revisions shows Wilde's development of formal modes which fuse subjective and objective expression into what he calls "art-criticism." Wilde's creation of criticism that is art, and art that is criticism, was his attempt to symbolize in written form the idealist dialectical theory developed from his synthesis.

In our discussion of "The Soul of Man Under Socialism" and "A Chinese Sage" we consider the utopian social implications of Wilde's synthesis and explain the coherence of his cultural and political criticism. We also contrast Wilde's ideas with Matthew Arnold's earlier attempts to put cultural criticism on a scientific basis.

The last section extends our analysis to include Wilde's critical fiction, "The Portrait of Mr. W. H." and *The Picture of Dorian Gray.* We argue that "The Portrait of Mr. W. H." develops a Hegelian interpretation of Shakespeare and his sonnets as the essential expression of the Renaissance spirit, that it fictionalizes the development of the critical and historical method Wilde described in "The Rise of Historical Criticism," and that it dramatizes the dialectical process involved in the perception and understanding of art. The three major characters in the story enact methods of interpretation which correspond to the stages of critical development Wilde describes in the "Rise" and "The Critic as Artist."

We interpret *The Picture of Dorian Gray* as a criticism of the materialist interpretation of ethics (viz. the pursuit of the greatest pleasure for the self) offered by Lord Henry Wotton. Instead Wilde affirms the reality and power of the spirit or soul in both art and criticism, a belief based on his acceptance of an idealist theory of art, the inheritance of acquired characteristics, and on W. K. Clifford's theory (recorded in the notebooks) that individual conscience is a manifestation of a powerful "tribal self."

In sum, our study presents a radical reinterpretation of Wilde's work based on our reconstruction, from the notebooks, of his synthesis and critical principles. We show the development of Wilde's dialectical understanding of history, criticism, and art from "The Rise of Historical Criticism" and his lectures in America to its fullest expression in *Intentions,* "The Portrait of Mr. W. H.," "The Soul of Man Under Socialism," and *The Picture of Dorian Gray*. We also provide an explanation for the formal innovations in his later criticism and fiction, namely, his fusion of subjective and objective modes, creation and criticism, in *Intentions,* "The Portrait of Mr. W. H.," and *The Picture of Dorian Gray*.

While Wilde's creative and critical achievements were unique, his education and general interests—his intellectual formation—were typical of the small group of serious Oxford students of the time. For this reason the notebooks and commentary provide a representative case study in late-Victorian thought. The notebooks also demonstrate the powerful synthetic impetus of humanistic education at Oxford and the consequent integration of scientific, philosophical, political, and aesthetic theories characteristic of this time. Wilde's readings and interpretations of major philosophers, literary figures, scientists, and social scientists offer a corrective to the reduced significance which they now have in histories of thought.

Our book title points to important aspects of this work. The main title is self-evident: it simply identifies the documents which are the object of our analysis and the basis for our reinterpretation of Wilde's work. The subtitle identifies three foci of our commentary. "Mind in the Making" suggests (1) Wilde's early education, his mind in the process of formation; (2) his interest in those aspects of Herbert Spencer's evolutionary theory and Hegel's philsophy which are concerned with the development of mind; and (3) Wilde's mature mind as it shaped his own literary criticism and fiction. We use the word "Portrait" because we recognize and acknowledge the interpretive nature of our scholarly and critical work.

Pittsburgh, Pa. P. E. S.
August 1988 M. S. H.

Acknowledgments

We are grateful to Mr. Merlin Holland, owner of the literary rights, for giving us permission to edit and publish these notebooks. We are also grateful to the William Andrews Clark Memorial Library of U.C.L.A., the owner of the manuscripts, for permission to publish and for assistance in research. Thanks to Edwin Floyd and Robert Marshall for help in translation and identification of classical sources. The Graduate School of the University of Pittsburgh financially supported part-time summer research assistants Margaret Wyda and Sean Monagle, who checked and proofread sources for Wilde's entries. Mary Briscoe, our Chair, generously granted us course reductions in support of research. John Bell of Computing and Information Services cheerfully provided assistance with countless problems involving computer software and hardware.

Our thanks to the anonymous reader at Oxford University Press for advice and criticism, and to our manuscript editor, Henry Krawitz, for correcting our errors of style and substance. Several other readers of several versions and portions of the manuscript made valuable suggestions, including Bruce Bashford, David Bartholomae, Janet Helfand, Robert Hinman, Harry Mooney, Thomas Philbrick, William Scheick, Susan Harris Smith, Richard Tobias, Frank M. Turner, Robert Whitman, and Phillip Wion. Stephen Regan helped identify Baudelaire and Swinburne as sources for remarks in the notebooks. Nicholas Rescher located a passage in Leibniz which Wilde quotes. Sir Rupert Hart-Davis and H. Montgomery Hyde patiently answered queries about the location of manuscript materials. J. E. Chamberlin gave us several productive suggestions about our research into the late nineteenth-century intellectual background. We are responsible for all remaining errors or omissions in research.

Individual Acknowledgments: Philip Smith

The University of Pittsburgh awarded me a Third Term Research Grant, an RDF Grant from the University Research Council, and a sabbatical leave, all of which gave me time and covered travel expenses for my research. Dong Xiangxiao, my graduate assistant funded by the RDF Grant during 1985–86, meticulously checked the quotations from secondary sources and performed invaluable services as a copy editor. Sharon Smartt provided careful and critical assitance as a proofreader of the notebooks at the William Andrews Clark Memorial Library.

The staffs of several libraries assisted generously. Since the beginning of the project, Hillman Library at the University of Pittsburgh provided me with a faculty study and excellent interlibrary loan service. Linda Delowry-Fryman, English department bibliographer, helped with research and materials. The William Andrews Clark Memorial Library of U.C.L.A. accommodated me several times since my first visit in 1976, when I found that they owned the manuscripts. Thanks to Thomas Wright, Carol Briggs, John Bidwell, and William Conway for providing study space, bibliographic information, photocopies of many documents, and good fellowship. The staffs of the Bodleian Library of Oxford University and the libraries of Clare and Magdalen Colleges were especially helpful during my sabbatical visit. Other collections useful to my research were: the Library of Congress; the Newberry Library and the John Crerar Library, Chicago; the New York Public Library, Lincoln Center; the British Library; the libraries of Northwestern University, the University of Illinois, Princeton University, Yale University, the University of Vermont, Carnegie-Mellon University, and the Pittsburgh Theological Seminary. Viscountess Eccles kindly permitted me to inspect one of Wilde's notebooks.

Thanks to Michael Helfand, especially for his unstinting work in the years 1976–78, when we were both judged for tenure in a university system unfamiliar with and unsympathetic to collaborative work in the humanities, and suspicious of the value of the lengthy research involved in our project. Our departmental colleagues, led by two chairs, Robert Hinman and Mary Briscoe, were unwavering in their support during this ordeal by bureaucracy.

My teachers Richard Ellmann, Frederic Faverty, and Gerald Graff generously shared their knowledge of Wilde and the nineteenth century and offered critical support for the project. Other teachers gave me more than they knew in ways they did not intend. I am grateful for the friendly but demanding encouragement, intellectual rigor, and exemplary commitment to teaching of John Darling, Donald Loftsgordon, Lewis Owen, Robert Ryf, Moody Prior, Walter Scott, and Bergen Evans.

This long project would have been more difficult without the support of parents, relatives, and friends who have also endured its going hence even as they did its coming hither. You know who you are and how much your encouragement mattered. Finally, loving thanks to Susan Harris Smith: for whom, everything; without whom, nothing.

Individual Acknowledgments: Michael Helfand

In the process of completing this long project, I received help from many institutions and individuals, and I am happy to acknowledge that support here. The University of Pittsburgh awarded me a summer research grant which gave me the opportunity to explore the intellectual wealth of *The Fortnightly Review*. The Hillman and Darlington libraries at the university, the Carnegie Public Library, Carnegie-Mellon University, and the Pittsburgh Theological Seminary all provided research materials crucial to the project, as did the library of Brown University, which allowed me to spend a summer researching its rich collection.

I would also like to acknowledge the help of teachers, colleagues, and friends.

James Boren, who fits all three categories, tried his best to teach me to think and write clearly. Frederick P. W. McDowell taught enormously wide-ranging and stimulating courses in mid- and late-Victorian literature. Sherman Paul introduced me to English and American cultural criticism and to the work of Raymond Williams. In China Irving Thalberg and Mary Farquhar took time from their teaching or research to read substantial parts of the commentary and to offer helpful editorial advice. My collaborator, Phil Smith, assumed administrative stewardship of the project after the writing and research were substantially done, but with much necessary and tedious work remaining. His work enabled me to spend two years abroad as a Fulbright lecturer. My largest professional debt is to Robert Scholes, who taught me as an undergraduate and shaped my graduate education in important ways by developing, and encouraging me to enter, an interdisciplinary Ph.D. program in modern letters at the University of Iowa. He has continued to support and encourage my work since I earned that degree.

Finally, my daughter, Sacha Helfand, encouraged me to write with her oft-reiterated question, "How long have you been working on this project, Dad?" Janet Helfand listened for years, with patience and goodwill, to my thoughts on Wilde and the trials and pleasures of researching and writing. This was a service well above and beyond the call of duty. Thank you all.

A Note on Collaboration

Because our collaborative writing speaks with one voice but is the product of two minds, it is impossible to separate individual contributions from the whole. We have worked together on the subject of Oscar Wilde since 1973, when we team-taught a graduate seminar in Victorian literature and began research on Wilde's short fiction. As the scope and focus of the project evolved as a result of our discoveries of Wilde's notebooks and his interests in idealist philosophy, evolutionary theory, and Hellenism, we continued, in our joint research and writing, to stress Wilde's seriousness and coherence as a critical theorist and author. Each of us brought different expertise to the enterprise; sometimes our varying perspectives produced disagreement, but our respect for each other's positions produced, we believe, a stronger essay.

Our method of writing derives from discussion, which produced composition and revision. We have, quite literally, sat together and jointly spoken and written every sentence in the commentary. We have also shared in the research which informs the commentary and the annotations for the notebooks. At academic conferences we have shared the reading of portions of our work in progress.

We claim equal authorship and joint responsibility for our writing. The order of our names on the title page is not intended to suggest primary and secondary contributions.

Contents

Oscar Wilde's Oxford Notebooks

1

The Text

This is an edition of two holograph notebooks by Oscar Wilde now housed at the William Andrews Clark Memorial Library of the University of California at Los Angeles. The longer of the two Wilde has entitled "Commonplace Book." It is a manuscript notebook (10.75″ × 8.5″ × .75″, now kept in a red morocco slipcase) of 133 pages with 2 inserted pages of a smaller size. The other, a manuscript notebook (9″ × 7″ × 1″, bound in roan back and marbled boards) catalogued as the "Notebook Kept at Oxford Containing Entries Dealing Mostly with Philosophical, Historical, and Literary Subjects," is a manuscript of 84 pages.[4] Neither was intended for publication, and both contain mostly quotations and paraphrases of other writers, along with Wilde's own analytical and descriptive comments, jottings, and fragmentary drafts. These notebooks probably were among the manuscripts listed for sale as Lot Seventy in the catalogue of the auction following Wilde's trial in 1895: "Manuscripts, a parcel."[5]

We have adopted editorial policies based on some of the guidelines and rationale for a genetic text provided by the Modern Language Association and the Center for Scholarly Editions,[6] but by their standards this is not a "pure" genetic text representing as nearly as print allows every feature of the manuscript. We have attempted to compromise between the needs of scholars and general readers by preserving the fragments and discontinuities, additions and deletions, idiosyncrasies, informality, and personal "feel" of Wilde's language and layout while also presenting an easily comprehensible text. Occasionally we have added bracketed punctuation marks when such changes clarify phrasal or sentence boundaries. We have added editorial explanations or conjectures in brackets. While we have represented most of Wilde's words exactly as written, some consistently misspelled words and scribal errors in the manuscripts are corrected; for example, Wilde's script often shows "e" for "a" in words like "what" or "that." We have accepted the penciled pagination of both notebooks because there is no need to change it and scholars have already used it to refer to entries. Since Wilde's entries were often brief, some of them consisting of only a line or two per page, we have frequently grouped several manuscript pages on a single page of this edition. Manuscript pages are separated from one another by an ornament and by the original pagination printed in brackets.

We also reproduce much of the visual layout of each manuscript page, including indentations and the various markings, such as underlining, marginal crosses, or the Greek semicolon [•] that Wilde used for emphasis or internal divisions on the page. We do not preserve the original line lengths within paragraphs. On a few manuscript pages which contain widely separated short remarks, we maintain the separations while compressing the blank spaces.

Wilde customarily wrote on the pages on the right side of the open notebook, and reserved the pages on the left for additional related comments. This practice is clear from the pagination of essays and quotations which cover more than one manuscript page. Pages containing additional comments thus occasionally interrupt the flow of a single extended entry. For the purposes of clarity, we indicate editorially where these interruptions occur. We also omit all blank pages separating entries and the blank pages at the end of the notebooks.

All annotations are separated from and follow the text. The major purpose of our annotations is to identify, where possible, the sources which Wilde quotes and paraphrases and his later use of these notes in published works. Since many readers of these documents will be scholars with interdisciplinary interests in fields such as literature, philosophy, history, and classics, we assume broad but not intimate historical knowledge of the various persons and subjects Wilde studied. We provide translations of Greek and Latin passages in the text and briefly identify, gloss, or explain relatively obscure entries.

Since Wilde seldom specifically identified his sources, the search for them has entailed a long, tedious, and humbling process of scholarly detective work. We have read systematically where the manuscripts provided some hints. We have pursued our own hunches and the suggestions of other scholars. We have inspected the recommended reading lists for the examinations in the *Literae Humaniores* program (''Mods'' and ''Greats'') in which Wilde took first-class honors at Oxford, and consulted the letters, biographies, and auction catalogue of Wilde's library for other records of his reading. We have looked in other unpublished manuscripts for clues and read relevant works by Wilde's teachers and contemporaries. And much more. Scholarship is long, but life is short. We have not identified all of Wilde's sources and uses for the materials in the notebooks, and would appreciate and acknowledge any suggestions readers of this edition could provide.

One of our editorial problems has been to distinguish among confirmed, probable, and possible sources for Wilde's entries. Whenever we have found in the notebooks a direct or indirect quotation, paraphrase, or summary of a passage from another work with clear and definite correspondence of vocabulary and idea, we have labeled that work a ''confirmed source.'' When the correspondence is less definite, but there are indications of indebtedness, we have identified that work as a ''probable source'' and have explained our reasons in the notes. Finally, we have made some educated guesses which we label ''possible sources'' when no definite source has been identified, no contextual evidence exists, but when there is a general correspondence of ideas or some similarity in phrasing.

Another editorial problem is the accuracy of Wilde's direct quotations from confirmed sources. Even when he indicated direct quotation, Wilde often made slight changes in the text. Since we have identified the sources and locations of the original passages, and since almost all are available in university collections, we have in most cases decided not to reprint them. In order to provide as clear a text as possible, we have keyed our notes to Wilde's text by using elided passages rather than by adding superscript marks or numbers.

The extended title, "Notebook Kept at Oxford Containing Entries Dealing Mostly with Philosophical, Historical, and Literary Subjects," gives a general sense of the range of subject matter in both manuscripts. A more detailed list would also include anthropology, sociology, chemistry, biology, psychology, political science, mathematics, religion, aesthetics, and the classics. But modern divisions of knowledge should not blind us to the ease with which educated Victorians moved freely and assumed connections among subjects that were only beginning to be recognized as disciplines. Wilde's own interests in such diverse subjects and his attempts to synthesize them exemplify these habits of mind. The titles Wilde gives his notebook entries often identify his subjects in terms of relations: "Matter and Mind," "Use of the poetic faculty in science," "Survival of Fittest in thought," "Moral Chemistry," and "Social and bodily organism."

There are various reasons why Wilde would be interested in these subjects. He was required to study some for university examinations and for fellowship competitions.[7] He was drawn to some by the practical considerations of writing "The Rise of Historical Criticism." In keeping a Commonplace Book, he probably intended to collect thoughts and quotations on significant and controversial subjects of the times.

These purposes suggest that the dating of both manuscripts needs to be modified.[8] The suggested beginning date for the Notebook Kept at Oxford (1874) is the earliest reasonable time Wilde could have started keeping the two notebooks; Wilde's Demyship at Magdalen College began in October of that year. There is substantial evidence which suggests that the concluding date for the College Notebook and the range of dates for the Commonplace Book are in error. The notebook sources with the latest publication date (1879) are T. H. Huxley's *Hume* and W. K. Clifford's *Lectures and Essays*, and there are quotations from several other works published after 1876.

We can also date the primary use of the College Notebook because much of the subject matter is relevant to, and several of the entries are incorporated in, "The Rise of Historical Criticism," which Wilde entered in the competition for the Chancellor's English Essay Prize of 1879. Wilde must have compiled and used these entries during the year that elapsed between the summer of 1878, when the topic for the competition ("Historical Criticism Among the Ancients") was announced, and the following summer, when the jury voted to make no award.[9] Of course, this does not prove that Wilde ceased using the College Notebook after that date. Although there are entries related to the essay throughout the notebook, there are also entries like those in the Commonplace Book concerning theoretical questions of science, history, philosophy, and art, subjects which Wilde continued to think and write about.[10]

Dating the Commonplace Book is more problematic. Wilde could have begun it as early as 1874, but he certainly kept it during 1878–79 since there are a few scattered entries which Wilde used when he wrote the "Rise."[11] While Wilde may have continued to keep the Commonplace Book during the 1880s, as the Finzi catalogue suggests,[12] we have so far no confirmed source later than 1879. However, we believe the influence of this manuscript cannot be determined by that last date since Wilde, like many writers, kept his Commonplace Book to preserve significant passages, important and shaping ideas, which he would later consult and review as he prepared to write. For example, Wilde's lecture given in America, "The English Renaissance in Art" (1882) contains passages and references which correspond to Commonplace Book entries, and in "The Portrait of Mr. W. H." (1889) Wilde's narrator recalls a Greek phrase from his Oxford days which appears as one of the "two Greek ideals" in the Commonplace Book (CW, 1195; CB, 40).

2

The Context of the Text

> I was a man who stood in symbolic relations to the art and culture of my age. I had realised this for myself at the very dawn of my manhood, and had forced my age to realise it afterwards.
>
> —*De Profundis*[13]

Until recently these notebooks have been dismissed as an interesting jumble of seemingly unrelated entries. We will argue that they have coherence and importance within the historical context of the philosophical, scientific, and literary movements that Wilde followed at the time. They serve as a map which charts his intellectual interests. For example, Wilde defines several topics, such as "mind and matter" or "historical methodology," in the notebooks. By tracing the various entries under these topics to their sources, we can, through contextual analysis, determine what he found relevant. This procedure leads back to the wider intellectual controversies Wilde confronted. By their very informality the notebooks are invaluable in further determining his intellectual position. They are like a compass which pointed Wilde (and so can point the reader) in directions which his later work followed.

Another context for the notebooks is Wilde's essay "The Rise of Historical Criticism." Because Wilde used the Commonplace Book and kept the College Notebook to write the "Rise," we use the published essay to analyze issues and questions he studied in the notebooks, as well as his conclusions. In the "Rise" Wilde addressed the assigned topic "Historical Criticism Among the Ancients" by considering the function of criticism through a historical critique of historical criticism. We will analyze this essay later in more detail in "The Text as Context," but because of the many relationships among Wilde's education, the essay, and the notebooks, we will introduce here some of the common themes. Most important are the origin of the critical spirit in the Hellenic race, the dialectical realization of this spirit in early historical and philosophical writings, and the systematic practice and theory of later Greek and Roman writers like Herodotus, Thucydides, Plato, Aristotle, Polybius, and Plutarch. As Wilde describes "the nature of the spirit of historical criticism itself in its ultimate development," he uses a Hegelian historical method to inform an analysis which "is not merely confined to the empirical method of ascertaining whether an event happened or not, but is concerned also with the investigation into the causes of events, the general relations which phenomena of life hold to one another, and in its ultimate development passes into the wider question of the philosophy of history" (CW, 1106).

Wilde organizes his essay on the principle of a dialectical development from unconscious motive to self-conscious awareness of criticism, from an instinctive, skeptical attack on religion and mythology by early Greek writers to Aristotle's rational development of a deductive-inductive historical methodology. He claims that the historical spirit can be best understood chronologically because Greek "intellectual development, not merely in the question of historical criticism, but in their art, their poetry and their philosophy, seems so essentially normal . . . so peculiarly rational, that in following in the footsteps of time we shall really be progressing in the order sanctioned by reason" (CW, 1106).[14] Wilde claims that Greek historiography anticipates its modern counterpart; except for a few modern contributions, "we have added no new canon or method to the science of historical criticism. Across the drear waste of a thousand years the Greek and the modern spirit joins hands" (CW, 1148).

Wilde offers Polybius as the historian who synthesizes in practice the theories Aristotle and Plato expounded: "He is the culmination of the rational progression of Dialectic" (CW, 1133). Wilde appreciates Polybius' uses of such modern historical assumptions as the Spencerian organicist notion of the Instability of the Homogeneous and the principle of heredity in organic life. Polybius also practices "modern methods of investigation" which require the empirical development of evolutionary norms as a basis for analyzing "social and political truths" (CW, 1136). Wilde ends the essay with a brief and dismissive account of Roman philosophers and historians, remarking that "Italy was not a pioneer in intellectual progress, nor a motive power in the evolution of thought" (CW, 1147). By minimizing Roman contributions and insisting on the unity of the Greek and modern spirits, Wilde consolidated the evolutionary idealist position which informs the essay.

Wilde's Irish Education

With this brief summary of the "Rise" as a guide to Wilde's ideas and interests, we can broaden the context for the notebooks by identifying the figures in Wilde's early life and education who shared and might have shaped his interests, opinions, and attitudes. Wilde's parents and their intellectual circle, as well as his education at Trinity College, Dublin, exposed him to ideas which appear both directly and indirectly in the notebooks and the essay.

He profited from the diverse but complementary scientific interests of his parents. Sir William and Lady Wilde studied and wrote about philological, ethnological, and archeological sciences, which they believed were crucial for social and political improvement. Sir William Wilde, renowned as a physician and statistician, "was one of the first to undertake the study of [the prehistoric Irish] in a precise and scientific manner."[15] He belonged to the Royal Irish Academy and the Celtic and Archeological societies: "[H]is three-volume catalogue of the contents of the Royal Irish Academy has been described as a monumental work of archaeological erudition and insight. . . ."[16] Among his acquaintances were the important archeologists John Gilbert and Sir George Petrie and their staff; William Rowan Hamilton, a mathematician, scientist, Astronomer Royal for Ireland, and classical and oriental philologist; and numerous others who made the Wilde's Merrion Square residence an intellectual center. Lady Wilde, known in the 1840s as the radical political writer "Speranza," shared her husband's interests and collaborated on some of Sir William's philological research; after his death she published essays on social, political, and literary subjects. Both parents, in differing ways, supported Irish political nationalism and the Celtic cultural revival. This family environment exposed Wilde to scientific, historical, cultural, and political analyses of modern and ancient societies.[17]

It is no surprise that Wilde, coming from such an intellectual environment, distinguished himself at Trinity College, Dublin. He belonged to the College Historical Society and the University Philosophical Society, where scientific and literary subjects were studied and debated. For work in the classics with J. P. Mahaffy he won the Berkeley Gold Medal and other awards at Trinity, and a Classical Demyship to Magdalen College, Oxford. Mahaffy's influence was both scholarly and personal: He helped guide Wilde's education in Greek language and culture and employed him as an editor for one of his books, *Social Life in Greece from Homer to Menander* (1874). After Wilde had gone to Oxford, Mahaffy included his ex-student on trips to Italy in 1875 and to Greece in 1877. The latter included visits to the excavations of Olympia and Mycenae, which provided Wilde with firsthand observation of the importance of archeological methodology in the study of Greece.[18]

Mahaffy was also a humanist who wrote widely and frequently on history, philosophy, mythology, anthropology, and philology. Even if the teacher and his pupil finally held different positions about certain theoretical problems, Mahaffy must have introduced many of the subjects and issues which Wilde takes up in the notebooks. For instance, in his *Twelve Lectures on Primitive Civilizations* (1869) and *Prolegomena to Ancient History* (1871) Mahaffy analyzes the practice of historical writing, the qualities of a good historian, and discusses the relative strengths of the "permanent causes"—language, geography, and race—which shape history. He describes and contrasts the methodologies developed by idealist philologists and materialist historians to interpret mythology.

As a member of the Church of England and a Kantian idealist, Mahaffy sought and often found the *via media* in these controversies. In his analysis of "permanent causes" in history, he introduces H. T. Buckle's materialist environmentalism as an alternative to the generally held view that race alone could explain national differences. He reconciles materialist and racialist theories by asserting the predominance of physical causes in prehistory and of racialist causes in civilized societies. He asks rhetorically, "[W]as there ever a civilization purely material? . . . Consider the nobler Semites and the Greeks . . . can any honest inquirer deny that here, indeed, are phenomena dependent not upon brute causes, but upon spiritual agents—not upon Matter, but upon Mind?''[19]

He takes a similar position in the controversy between historical and philological approaches to myth. In ''The Value of Legends and Critical History,'' Mahaffy questions the materialist historicism of George Grote's approach to myth, but he also cautions against uncritical acceptance of the idealist methods of comparative mythology. Mahaffy's position, finally, is closer to F. Max Müller's philological idealism, which does not reject the possibility of historical sources but still finds the origins of myth in primitive imagination and language: ''It will make a close attention to similarity of names in both sound and etymology, of absolute necessity in convincing us of the single origin of parallel legends. But even beyond this, it will open to us a new page in the forgotten life of pre-historical ages, and will show us that, as every other part of our nature obeys its laws, so even the riotous faculty of the imagination is checked by some secret rein and guided unconsciously upon beaten paths.''[20]

This belief takes on added significance when Mahaffy discusses the nature of history and the qualities of a good historian. True history, he says, is more than a collection of facts about kings, battles, and sieges. Historical materials required analysis and appreciation informed by a ''historic sense.'' This historic sense is a ''quality inborn in some men . . . . It is of course trained, improved, and corrected by study and experience, but no ordinary man will ever create it within himself . . . the faculty is a sort of instinct.''[21] He defines this historic sense as a combination of ''ruthless skepticism and a vivid imagination,'' qualities which, he insists, are not contrary but complementary and ''constantly combined in great historians.''[22]

Wilde at Oxford: The Influence of F. Max Müller

Wilde's study with Mahaffy was the capstone of his educational experiences in Ireland; he arrived at Oxford in 1874 with an extensive background in the histories of classical and Irish civilizations and an introduction to the problems and controversies surrounding historical interpretations. ''My dear Oscar,'' his teacher is said to have remarked, ''you are not clever enough for us in Dublin. You had better run over to Oxford.''[23] While it is doubtful that Oxford made Wilde cleverer, it certainly furthered his education in significant ways. For his degree Wilde studied ancient and modern philosophy (i.e., ethics, metaphysics, logic, and political theory), philology, divinity, the classics, and history. The most important influences from this program of study came in philosophy, where Wilde adopted Oxford Hegelianism, and in philology, which he studied with F. Max Müller. In both philological and philosophical studies, Wilde learned of the possibility of reconciling evolutionary science with philosophical idealism. The faculty at Oxford made the curriculum of *Literae Humaniores* relevant by relating it to modern controversies. Like many Victorian teachers, they examined the past from modern analytical perspectives, often turning it into a mirror for contemporary debates over theology, aesthetics, politics, and philosophy.[24] In the 1870s theories of human evolution were central to controversies in scientific, social, and humanistic disciplines; philologists drew upon approaches from all three areas.

The most eminent comparative philologist at Oxford, indeed in all of England, was Müller, whom Wilde knew and studied with during his first year. "Max Müller loves him," Lady Wilde wrote to a friend in 1875.[25] Müller, educated in Germany in the 1840s, brought to Oxford his historical research in comparative languages, religions, and mythologies, as well as a belief in philosophical idealism that grew out of his German university experience and was partly based on an extensive knowledge of Hegel's work.[26] During the controversy over Darwin's theory of natural selection, Müller found himself in an ambivalent position. He had used the theory to explain the growth of Western languages and had accepted Darwin's explanation of the development of nonhuman species. But he differed with Darwin's materialist explanation of human nature and origins. Language, which was inseparable for Müller from all mental activity, was "the one great barrier between the brute and man: Man speaks, and no brute has ever uttered a word. Language is our Rubicon, and no brute will dare to cross it. . . . Language is something more palpable than a fold of the brain, or an angle of the skull. It admits of no cavilling, and no process of natural selection will ever distill significant words out of the notes of birds or the cries of beasts."[27]

After tracing the elements of all languages to their four to five hundred original "phonetic types," Müller asked, "How can sound express thought? How did roots become the signs of general ideas?" He answered, "They are *phonetic types* produced by a power inherent in human nature. They exist, as Plato would say, by nature; though with Plato we should add that, when we say by nature, we mean by the hand of God" [emphasis in original] (SL, I, 384).

Müller's idealism also informed his comparative philological method and his histories of linguistic origins and development. Based on his study of the earliest Indo-European languages, including Greek and Sanscrit, he claimed that a lost parent language was created by divinely inspired, imaginative, and rational members of the Aryan race. The original language was then transformed by racially related nomadic groups into later languages and mythologies, fallen versions of the original tongue. Thus, he called mythology a "disease of language" (SL, I, 21). This "golden age" theory of language had significant implications for ethnologists, etymologists, and interpreters of myths. It gave support to Wilde's belief that the critical spirit was a racial characteristic originating with the Aryan Greeks. For etymologists, and for Müller's acquaintance John Ruskin, the theory implied that the true meaning of any word could be determined by finding its root, since that root was the purest expression of the language.

Müller's comparative linguistic analysis had considerable impact on the study of philology and myth in the second half of the nineteenth century, constituting "probably the most influential academic view of myth to appear before the turn to anthropological theory,"[28] even though many of his assumptions, conclusions, and methods were challenged and proven erroneous. Müller used this academic prestige and specialized knowledge to defend religious belief and reconcile it with Darwin's materialist explanation for nonhuman life.

This reconciliation, which seems so contradictory and strange today, was quite typical of a *via media* reached by mid-century religious and secular intellectuals. Two historians that figure significantly in Wilde's notebooks, W. E. H. Lecky and H. T. Buckle, are typical of rationalist, secular thinkers in accommodating the possibility of God, the spiritual, and the soul with practical assumptions that deny divine intervention in history. Noel Annan suggests that, unlike their continental counterparts, these rationalists—along with more famous ones like Herbert Spencer, T. H. Huxley, and W. K. Clifford—refused to push materialism to its logical conclusions for political and social reasons.[29]

Wilde at Oxford: The Influence of John Ruskin

When Wilde studied with Müller, he also met John Ruskin, then Slade Professor of Fine Arts. Ruskin's well-known influence on Wilde, we will suggest, extends beyond art and aesthetics: The College Notebook shows that Wilde read *The Queen of the Air* (dealing with the myth of Athena) and was particularly interested in the relationship of the imagination to science, mythology, and poetry. In that study Ruskin acknowledged the importance of Müller's philological and mythological theories in reshaping his own method of interpretation.[30] An idealist like Müller, Ruskin freely adapted the findings and hypotheses of modern science but opposed its materialism. He believed that the origin of myths could be found in historical events or natural phenomena like sun, sky, cloud, or sea.

While "the root, in physical existence," might be a helpful warrant for interpreting the meaning of a myth, Ruskin disagreed with Müller about the root as real meaning and about mythology as a disease of language; instead he found the "real meaning" of a myth "at the noblest age of the nation among whom it is current. The farther back you pierce, the less significance you will find, until you come to the first narrow thought, which, indeed, contains the germ of the accomplished tradition; but only as the seed contains the flower."[31] Ruskin, like Müller, did not believe that the meaning of a word or a myth was readily available to any reader or, indeed, to its writers or originators. To create or interpret great myths both the poet and the reader must go beyond "frigid scholarship" and have "true imaginative vision" because it alone provides truth which is "vital, not formal . . . and it is this veracity of vision [of the men who tell myths] that could not be refused, and of moral that could not be foreseen, which in modern historical inquiry has been left wholly out of account: being indeed the thing which no merely historical investigator can understand, or even believe; for it belongs exclusively to the creative or artistic group of men, and can only be interpreted by those of their race, who themselves in some measure also see visions and dream dreams" (QA, 309). "Right reading of myths" requires a relationship between an imaginative writer's text and an imaginative reader, "the understanding of the nature of all true vision by noble persons. . . . [It] is founded on constant laws common to all human nature; that it perceives, however darkly, things which are for all ages true;—that we can only understand it so far as we have some perception of the same truth;—and that its fulness is developed and manifested more and more by the reverberation of it from minds of the same mirror-temper, in succeeding ages'' (QA, 310).

Ruskin's emphasis on the importance of vital, imaginative, and spiritual truth appears even in his choice of myth. According to Ruskin, in two of her three manifestations Athena represents the power of spirit: ''[S]he is the Spirit of Life in material organism; not strength in the blood only, but formative energy in the clay: and, secondly, she is inspired and impulsive wisdom in human conduct and human art, giving the instinct of infallible decision, and of faultless invention'' (QA, 346). Athena as spirit, he adds, is analogous to ''the ministry of another Spirit whom we also, believing in as the universal power of life, are forbidden, at our worst peril, to quench or to grieve'' (QA, 346).

Ruskin saw no contradiction between his spiritual and vitalist interpretation of life and the scientific explanations offered by Darwin, Huxley, and John Tyndall.[32] His description of section II of *The Queen of the Air*—''Athena Keramatis,'' as a ''Study . . . of the supposed, and actual, relations of Athena to the vital force in material organism'' (QA, 351)—allies him to one camp, the vitalists, in the ongoing debate within the scientific community concerning the origin and nature of life. Many scientists, like Ruskin, thought that ''life really was something that acted through the medium of matter but was independent of it; before matter could assume the attributes of life it was necessary that a 'vital force' be added to it.''[33] For Ruskin the spirit of life informed plants, all species of animals, and man alike, giving shape and form to matter. So, for example, he says, ''The Spirit in the plant . . . [is] its power of gathering dead matter out of the wreck round it, and shaping it into its own chosen shape'' (QA, 357). For humans this vital force, present in the powers of imagination, morality, and the aesthetic sense, recognizes and appreciates the workings of the spirit in both natural and human creations, and itself is capable of creativity.

Ruskin had a vitalist explanation, again adapting a current scientific theory of how spirit informs nonliving matter, and how it enables spirit in one living form to relate to spirit in another. Briefly, Ruskin thought that the spirit of life resided in the heat and motion which animated and informed matter and gave it life through vibrations, which moved through a medium called ether. These vibrations affected all living things and in humans were the physical means not only of perception but also of that aesthetic and moral capacity to see spirit in nature (QA, 292, 356, 378).

For Ruskin nature contained moral meanings, and humans, provided they had sufficient individual and racially inherited moral character and aesthetic sense, could recognize these meanings and re-create them in artworks. Ruskin's fundamental principle was that ''art is the work of the whole spirit of man; and as that spirit is, so is the deed of it: and by whatever power of vice or virtue any art is produced, the same vice or virtue it reproduces or teaches'' (QA, 391). Human abilities and conduct were a matter of free choice and were ''the result of the moral character of generations. . . . Perseverance in rightness of human conduct, renders, after a certain number of generations, human art possible . . . '' (QA, 393). Free will and inheritance, as causes of moral character and what Ruskin called the ''inherited art-gift'' (QA, 396), were fully compatible with and scientifically justified by Darwin's and Herbert Spencer's racialist theory of the inheritance of acquired characteristics.[34] Ruskin cited himself as an example: "This art-gift of mine could not have been won by any work, or by any conduct; it belongs to me by birthright, and came by Athena's will, from the air of English country villages, and Scottish hills" (QA, 396).[35]

In his College Notebook Wilde chose two examples from *The Queen of the Air* to illustrate his entry "use of the poetic faculty in science" (CN, 42–44). While he mentions several ways in which the poetic imagination apprehends scientific truths, he applies Ruskin specifically to the statement that the early Greeks "had mystic anticipations of nearly all great modern scientific truths" (CN, 43). Understood in their larger context, these notes indicate that Wilde accepted Ruskin's philological method for interpreting mythology, as well as some of the assumptions which inform that method. For instance, when he says, "So Ruskin shows the inherent truth in ancient symbolism" (CN, 42), Wilde reveals his idealist assumptions about the nature of truth (it exists in germ at the beginning of history) and its development in history (it is realized more and more self-consciously in time); more specifically, he reveals his idealist assumptions about mythology (there is truth in it not in its historic origins but in its imaginative foreshadowing, and an imaginative philological interpretation must be used to learn the truth).

Wilde would also have been exposed to Ruskin's idealist theories of history, perception, art, and science when he attended the lectures on "The Aesthetic and Mathematical Schools of Art in Florence" during Michaelmas Term (10 November to 14 December 1874).[36] Ruskin interpreted the historical development and decay of Florentine art from the thirteenth through the fifteenth centuries by using an idealist assumption of the unity of moral and aesthetic faculties and a racialist explanation for their origins. He identified five stages of Florentine art: "(1) Savage; (2) AEsthetic; (3) Mathematic; (4) Attempt to regain the AEsthetic, which may be called Romantic, combining faith and imagination against (5) Sensual and Infidel" (AM, 186n). Ruskin's idealist theory of art and perception may be found in his distinction between two "tempers" or states of mind, aesthesis and mathesis (AM, 211). When aesthesis dominates, artists see spiritually and create works which embody spiritual truth; when mathesis dominates, artists neglect spiritual truth and strive for measured or mathematical perfection of form.

In the history of Florentine art the "Aesthetic" school of 1300 produced the greatest art because it saw spiritually: "To the school of Perception—that which depends on its instinctive sight and sense—belongs necessarily the foundational discovery of the existence and true nature of things; while to the demonstrative, instructive, or mathematical school [of 1400] belongs the comparison, discipline, arrangement, and correction of impressions received by the senses" (AM, 249). According to Ruskin, the secondary artistic achievement of the mathematical school was caused by a diminution of faith, a consequent loss of imagination, and a greater interest in "material objects" and "abstract theorems" (AM, 185–86, 212, 214).

Ruskin explained that the faith, imagination, and moral power of the Aesthetic school developed because of the union of northern and southern races in Florence.

He extends his racialist explanation when he says of Cimabue and Giotto, "My own belief is that these men are both absolutely of Graeco-Etruscan race, as opposed to the Norman; that they represent the new budding of an underground stem which has its root partly in Greece proper, partly in Egypt. . . . They are at once Greek of the Greeks, and Christian of the Christians—the flower and purest force of both" (AM, 200). This racialist theory is no metaphor. Ruskin believed that the inheritance of acquired characteristics explained the physical differentiation of the races and, in this case, the development of an instinctive aesthetic sense.[37] He alludes to this process, near the end of these lectures, when he begins to suggest the implications for modern society of his historical and aesthetic analysis. Ruskin admits that even if some modern scientists and physicians deny it, there is a science of religious and spiritual life. He reconciles his idealist theory of art with his acceptance of contemporary science through his use of the inheritance of acquired characteristics. Qualities of mind and character are passed on biologically, he says, and they are accompanied by physiological changes:

> For the characters and passions of men descend and proceed from each other as trees do from graft or seed; there is a botany, a science, of the growth of the mind which lets you see either intellect or conscience unfolding, first the blade, then the ear, after that, the full corn in the ear. Parallel with these mental changes there are changes in our body and in the nervous substance of the brain; so that an Etruscan brain would differ from a Gothic one, and Quercia's from Ghiberti's, in entirely physical particulars. (AM, 263; see CB, 83)

Ruskin assumes that materialist science denies the reality of spirit and that this assumption blinds those who hold it to manifestations of spirit in life. This blindness, in turn, can lead, through custom and practice, to a biological inability of an entire race and culture to see aesthetically and morally (i.e., idealistically)—especially in England: "And yet people speak in this working age, when they speak from their hearts, as if houses and lands, and food and raiment were alone useful, and as if Sight, Thought, and Admiration were all profitless, so that men insolently call themselves Utilitarians, who would turn, if they had their way, themselves and their race into vegetables . . ." (MP II, *Works* IV, 29).

On the other hand, through education and practice an idealist instinct for beauty and good conduct could be strengthened and could help an entire racial group to see better, as Ruskin's famous remark in *Modern Painters* suggests: "[T]he greatest thing a human soul ever does in this world is to *see* something, and tell what it *saw* in a plain way. Hundreds of people can talk for one who can think, but thousands can think for one who can see. To see clearly is poetry, prophecy, and religion,—all in one" (MP III, *Works* V, 333). Wilde knew *Modern Painters*;[38] in fact, Ruskin gave a series of lectures at Oxford in 1877 which included readings from the second volume that outlined his idealist theory of true and false artistic creation and aesthetic perception. We argue that some of Ruskin's critical interests and assumptions decisively influenced Wilde's development. He shared with Ruskin an idealist perspective and consequently an antimaterialist and antiutilitarian position in culture and politics. He also shared an interest in how the imagination functioned to create and interpret art and to improve culture.

Wilde at Oxford: Ruskin's Idealism Versus Pater's Materialism

In his 1877 lectures Ruskin attacked the growing "folly" of the Aesthetic movement, which, he believed, based itself upon materialist philosophy ("Introduction," MP, *Works* IV, 35). These lectures were the last in a series (begun in 1870, when Ruskin assumed the Slade Chair) in which he warned against the "corruption" of art by the "materialism of the art for art's sake movement," which degraded appreciation to "a mere operation of sense, or perhaps worse, of custom" (MP II, *Works* IV, 35–36).[39] Ruskin's attack on James McNeill Whistler as an aesthete is well known. At Oxford, however, the serpent in the bosom of moral art was Walter Pater, whose *The Renaissance* (1873) represented a decisive theoretical turn against Ruskin and idealist aesthetic theories. Both Richard Ellmann and Michael Levey describe the intellectual rivalry between Pater and Ruskin, interpreting *The Renaissance* as an implicit challenge to Ruskin's theories of art and culture.[40]

This rivalry over aesthetics was rooted in differing interpretations of Aristotle's *Ethics*, which was "the single most translated Greek philosophical text of the entire century," and was then required reading in the *Literae Humaniores* program at Oxford (Turner, pp. 323–24).[41] Pater, in his "Conclusion," makes a materialist, relativist, and subjectivist argument which was heavily indebted to Sir Alexander Grant's mid-Victorian historicist interpretation of the *Ethics*. Pater adopts Grant's position that *energeia*, or heightened consciousness, is more important than conduct in Aristotle's philosophy:

> Pater's emphasis on the pursuit of moments of pleasurable experience and contemplation of them from an aesthetic rather than a moral point of view were congruent with Grant's analysis of Aristotle's use of *energeia*. . . . Pater's injunction to seek a variety of experiences evoked Grant's paraphrase of the meaning of Aristotelian pleasure as "the sense of life itself; the sense of the vividness of the vital powers; the sense that any faculty whatsoever has met its proper object." Both Grant and Pater believed such pleasure was applicable to the mind and the physical senses. . . . Throughout the "Conclusion" Pater set the same high value on the heightened state of consciousness as the source of both pleasure and happiness that Grant had argued was conveyed by a proper understanding of *energeia* in Aristotle's philosophy. . . . [Pater] had appropriated the parts of Grant's analysis and rhetoric that challenged contemporary morality rather than those that confirmed it. (Turner, p. 354)

This interpretation allows Pater to claim that the cultivation of consciousness—the maximizing of an individual's experience by being "present always at the focus where the greatest number of vital forces unite in their purest energy"—is "success in life." Because of his materialist perspective, Pater rejects "philosophical theories or ideas," any "facile orthodoxy, of Comte, or of Hegel, or of our own"; they have no ontological reality and are valued pragmatically only insofar as they "help us to gather up what might otherwise pass unregarded by us."[42]

His position becomes, then, a utilitarian argument for aesthetic hedonism based on individual contemplation and enjoyment of sensual experience. It shocked the late-Victorian art establishment, and Pater temporarily suppressed the "Conclusion." It also must have shocked Ruskin, whose *Modern Painters*, II, was originally written in the 1840s based on his earlier, Christian interpretation of Aristotle's *Ethics*. When Ruskin decided to lecture and later to revise *Modern Painters* in the late 1870s and early 1880s, his opposition to aestheticism was based on yet another turn in the interpretation of Aristotle, one which arose among "Idealist philosophers at Oxford [who] used the *Ethics* to renew the long-standing university opposition to political individualism and utilitarianism" (Turner, p. 358). This school of interpretation, as practiced by T. H. Green, "attempted to mesh the ethical concerns of the [earlier] Christian commentators with the historical outlook of Grant and with the social concerns of idealist philosophy" (Turner, p. 360).

Ruskin's revisions of *Modern Painters* reflect this idealist reinterpretation of the *Ethics*. He repudiates what Cook and Wedderburn call the "sectarian narrowness" of his 1840s Christian interpretation ("Introduction," MP II, *Works* IV, xlvii) and finds a new and wider significance in Aristotle's concept of *theoria*. In the 1883 "Preface to the Re-arranged Edition" Ruskin says he has taken "Theoria" (which he translates as "contemplation") from "that great sentence of Aristotle's [which] is the conclusion of all the moral philosophy then taught at our universities. . ." (MP II, *Works* IV, 7).[43]

In *Modern Painters* he defines two related faculties of mind, the "Theoretic" (contemplation as spiritually informed perception), and the "Imaginative" (which is divided into three operations: Penetrative, Associative, and Contemplative).[44] The Theoretic faculty "is concerned with the moral perception and appreciation of ideas of beauty" (MP II, *Works* IV, 35). He argues that this "exulting, reverent, and grateful perception" is necessary for "the full comprehension and contemplation of the Beautiful" (MP II, *Works* IV, 47), and contrasts this faculty with what he calls "AEsthetic" perception, or "aesthesis," which "properly signifies mere sensual perception of the outward qualities and necessary effects of bodies" (MP II, *Works* IV, 42).[45]

Ruskin describes the imagination as "the second great faculty . . . which the mind exercises in a certain mode of regarding or combining the ideas it has received from external nature, and the operations of which become in their turn objects of the theoretic faculty to other minds" (MP II, *Works* IV, 36). Like the Theoretic, the Imaginative faculty partakes of the spiritual and so perceives the spiritual and beautiful in objects. "Under the discipline of the theoretic" (MP II, *Works* IV, 314), it helps artists make artworks which other minds can contemplate. Ruskin's system of portraying the perception, conception, and creation of artworks, then, included the imaginative transformation of subject matter to reveal spiritual meaning and noble use. For Ruskin both contemplative and imaginative pursuits are ultimately useful, but he makes it quite clear that their usefulness resides in countering the utilitarians' sense of "use."

According to Ruskin, utilitarianism encourages individuals to perceive objects in the world materially, and "to discover of what selfish uses they are capable" (MP II, *Works* IV, 35). Theoretic and imaginative pursuits (such as science and the arts) understand and interpret things as ends in themselves and for their own sakes and stand against the selfish vision which, Ruskin believed, informed both men and nations (MP II, *Works* IV, 32–35). Ruskin's ideal was that life would be an art, and all humans would perceive and act unselfishly. "Utility" would have a humanistic rather than an economic meaning. Ruskin's political economy and conceptions of utility are often expressed in language full of scriptural allusions, but his ideas are based on a humanistic rather than a sectarian principle. The virtues found in the Bible and in "heathen literature" may imply the

> conception . . . of personal and governing Deity: but they begin, practically, and themselves consist to the end, in truthful knowledge of *human power* and *human worth*. . . . This character, intelligently obedient to a moral law common to the Jew and Arab,—to the Greek and Christian,—the past world, the present world, and the world to come,—is assumed here, and in all my other writings whatever, as the *basis* of religion itself,—not religion as the basis of *it*, and the first condition of true delight in the contemplation of any visible thing, or the conception of any invisible one; for only in this state of mind can we see that anything is Good, in the sense that its Creator pronounced it so [Emphasis in original]. (MP II, *Works* IV, 6)

Ruskin's assertion that a morally informed state of mind is the "basis" of the good life assumes an interpretation of Aristotle's *Ethics* which sees *theoria* rather than *energeia* at its core. The ability to contemplate, that is, to have true knowledge of human power and human worth is simultaneously the means to and the end of the good life. Although he was no Hegelian, Ruskin's interconnection of means and ends (i.e., his recognition of their dialectical relation) would have made his theory amenable and attractive to more self-consciously philosophical idealists like Oscar Wilde.

Wilde had also read and carefully studied the *Ethics*; in a short notebook he wrote an outline and commentary on Books VI and VII which contain borrowings from Sir Alexander Grant's footnotes.[46] He would therefore have been familiar with the differing interpretations of the *Ethics* and with the positions taken by Pater and Ruskin. In his Commonplace Book Wilde implicitly adopts the *theoria* interpretation. After he identifies contemplation as a good in itself ("an 'αρετη of the soul' "), he, like Ruskin, relies on the *Ethics* to make a distinction between theoretic and aesthetic seeing, between contemplation and mere sensual perception (CB, 189–91). Wilde is even more specific about the difference between contemplative knowledge and aesthetic perception. The latter is limited to the impressions of isolated facts or sensible things, whereas knowledge produced by contemplation can inform us of the "universality and necessity which are the factors as well as the product of experience," that which, as his Greek phrase suggests, is ever and everywhere true (CB, 191).

This comment follows a quotation from Hegel on philosophy's immediate lack of utility and puts Wilde (with Ruskin and Aristotle) among those who recognize contemplation as a good different from and superior to immediate moral actions such as philanthropy. Wilde cites Aristotle's disavowal of any philanthropic purpose for knowledge. The contemplative life is an end in itself and not an instrument to improve the world: "[I]ts duty is to comprehend the world not to make it better" (CB, 189). Wilde also distinguishes between seeing and true seeing, but unlike Ruskin he describes the importance of contemplation in Hegelian terms:

"Its' [*sic*] sphere is the Universal—The universal Law of the Minds [*sic*] movement[.] It examines not what ought to be but what is• It is the possession of real wisdom, not the love of wisdom•" (CB, 189).

Wilde's interpretation of Aristotle's *Ethics* and the importance he, like Ruskin, gave to contemplation (*theoria*) rather than Pater's *energeia* place him firmly in the camp of the Oxford Hegelians, who used the *Ethics* to counter utilitarianism and materialism. Philosophy, contemplation, or wisdom might have no specific philanthropic ends, as Wilde remarks, but, rightly understood, as Ruskin and the Oxford Hegelians might have said, they can have socially significant, even utopian, aims and consequences. "The Soul of Man Under Socialism" (1891), we will argue, proves Wilde's place in this tradition.

Wilde and the Oxford Hegelians

Wilde's use of philosophical terms shows the powerful influence of the Oxford Hegelians, whose ideas and writings he recorded at greater length than Ruskin's in the notebooks. The Commonplace Book contains many notes from William Wallace's "Prolegomena" to *The Logic of Hegel* (1874) and from Benjamin Jowett's Hegelian introductions to the second edition of *The Dialogues of Plato* (1875). These men and other Hegelians had such an impact on undergraduates that for those, like Wilde, who were "In the 'Greats'-circle, Philosophy was our mistress that claimed most of our devotion, roused most of our ardour, and decided our fate in the schools."[47] Like Müller and Ruskin, these idealists were well aware of the challenge Darwinian theory and materialism offered to traditional theology, philosophy, and morality. They also believed that Hegel's philosophy encompassed and reconciled evolutionary theory with a metaphysical explanation of human life. In fact, as J. E. Chamberlin has noted, "Many Oxford students in the 1870s claimed that for a long period they never saw Darwin except through Hegelian bifocals."[48] Wilde was one of them.

The Oxford Hegelians' rationale begins with a crucial problem in classical philosophy, namely, the relation of Being and Not-Being. They solved it by asserting the metaphysical reality of a world of flux, that is, by insisting that truth was a dialectical relation of Being and Not-Being. The theory of natural selection also posited a world of flux which evolved through a process of conflict. But, as William Wallace said and Wilde noted, "Modern science is aetiological . . ." (CB, 205). Hegelian philosophy encompassed evolutionary theory by providing a final cause or purpose: the evolution and revelation of truth or Mind through dialectical conflict. Evolution therefore became the law of both physical and mental development: The history of both Nature and Philosophy was dialectical and evolutionary.

Wilde develops his own synthesis by drawing upon Wallace's and Jowett's works. He borrows from Wallace's "Prolegomena" both a pervasive assumption—the similarity of the history of philosophy, or development of Mind, to the process of biological evolution—and a number of passages containing solutions to significant metaphysical problems. Wilde accepts Wallace's critique of two well-known first principles of philosophy: his Aristotelian refutation of the principle of utility (CB, 201), and his recognition of the inadequacy of Descartes' "*Cogito, ergo sum*" as a beginning for a philosophical system. Descartes' problem was, Wallace says, that he did not begin at the beginning:

> To begin with the "I" would only place us at a point where the severance between thought and being was already a *fait accompli*. . . . The beginning of philosophy or logic must go far deeper than this original division. We must see thought in its first and fundamental calling. As the biologist, when he describes the structure of a plant, begins on the assumption of a previous development of parts, with an existing plant, which has resulted in a seed,—but begins with the seed from which the plant is derived: so the logician must begin with a point which in a way pre-supposes the system to which it leads. . . . Thus the beginning of Logic presumes the existence of Absolute Mind, as the beginning of botany presumes the existence of the plant. It is from this circumstance that Hegel describes philosophy as a circle rounded in itself, where the end meets with the beginning, or says that philosophy has to comprehend its original grasp or concept.[49]

Wilde pursued two topics in Wallace's history of philosophy: his assertion that philosophy's dialectical development constituted natural selection in the world of thought, and his placement of ancient and modern philosophers in history. These topics figure crucially in Wallace's synthesis of Hegelian and evolutionary theories. For Wallace the progress of reason constituted a dialectical relation of creative, positive speculation and negative, destructive skepticism. In the "Prolegomena" he describes it as part of the process of evolution:

> Let us put the same law in a more concrete case. It is reason,—the Idea,—or, to give it an inadequate and abstract name, Natural Selection—which has created the several forms of the animal and vegetable world: it is reason, again, which in the struggle for existence contradicts the very inadequacies which it has brought into being: and it is reason, finally, which affirms both these actions,—the hereditary descent, and the adaptation—in the provisionally permanent and adequate forms which result from the struggle.
>
> The three stages thus enumerated are not merely stages in our human reason as subjective. They state the law of rational progress or development in pure thought, in Nature, and in the world of Mind,—the world of Art, Morals, and Science. They represent the law of thought or reason in its most general or abstract terms. ("Prolegomena," p. cii)

Furthermore, Wallace believed that the operation of the dialectic in the history of philosophy could be found by a method analogous to the one Darwin used to discover natural selection, namely, by a study of the evidence from artificial selection in the breeding of domestic animals: "[T]he history of philosophy is the conscious evolution of what for the far greater part is transacted in the silent workshops of nature. Philosophy, in short, is to the general growth of intelligence what artificial breeding is to the variation of species under natural conditions" ("Prolegomena," p. cx). Wallace also saw in the history of philosophy a common source of energy for mental and physical evolution. In his notes drawn from Wallace, Wilde recorded Plato's definition of the "dialectical energy of thought" as "Eros" and, on the facing page, he paraphrased Wallace: "Hegelian dialectic is the natural selection produced by a struggle for existence in world of thought" (CB, 204–5). He also noted Wallace's definition of the dialectic as "the spirit of skepticism—the force of contradiction and the negative" ("Prolegomena," p. cii).

Wallace's interpretation also informed Wilde's understanding of individual philosophers in this evolutionary dialectic. Wallace saw parallels between ancient Greeks and modern Europeans such as Socrates and Kant, Aristotle and Hegel. He characterized the history of philosophy as a rhythmic oscillation between idealist and realist schools and believed that traditional philosophical problems, such as the relation of mind and matter, reflected in their very definitions an incompleteness in both schools which could be resolved only by recognizing an underlying synthetic unity ("Prolegomena," pp. lix–lx). Wilde noted these ideas in the Commonplace Book and later incorporated them in "The Rise of Historical Criticism" (CB, 101, 105). In his entry "The Origin and progress of Ancient and modern philosophy," Wilde adopted examples from Wallace to illustrate parallels and progress in thought (CB, 209; see also CB, 111).

Wallace describes Hegel's solution to the fundamental metaphysical problem of the relation of Being to Not-Being: He denies the reality of the Cartesian dualism and posits instead an original and final unity. That is, he denies sole ontological status to the rock of "Ego," the metaphysical foundation of Descartes' system. Instead, metaphysical reality lies in relation:

> The subjective world—the mind of man—is really constituted by the same force as the objective world of nature: the latter has been translated from its externality of parts in time and space, into an inner world where unity, the fusion or coalescence of all types and forms, is a leading feature. The difficulty of passing from the world of being to the world of thoughts, from notion to thing, from subject to object, from Ego to Non-ego, is a difficulty which men have made for themselves in their theories. They reasoned on the ground that the individual mind was a fixed and absolute centre, from which the universe had to be evaluated. In Hegel's words, they made man and not God the object of their philosophy. So that Kant really showed the outcome of a system which acted on the hypothesis that man in his individual capacity was all in all. Hegel, on his own showing, came to prove that the real scope of philosophy was God,—that is, the "original synthetic unity" from which the external world and the Ego have issued by differentiation, and in which they return to unity. ("Prolegomena," p. lx)

In an entry entitled "Metaphysics" Wilde seized upon Wallace's Hegelian argument denying fundamental significance to the Ego, and hence to the individual; instead, reality must be understood as a dynamic relation between Self and Other (CB, 199). Wilde also compiled notes from an extended commentary on the subject in Benjamin Jowett's introductions to Plato's "Sophist" and "Parmenides" (CB, 111, 195). Jowett described the origins of the problem in Eleatic philosophy and outlined Plato's answer, which Jowett saw as an anticipation, however incomplete, of Hegel's dialectical solution. As Jowett noted, "the kindred spirit of Hegel seemed to find in the Sophist the crown and summit of Platonic philosophy—

here was the place at which Plato most nearly approached to the Hegelian identity of Being and not-Being.''[50]

Jowett elsewhere describes the problem posed as ''the crux of ancient thinkers in the age of Plato: How could one thing be or become another?'' (DP, IV, 401). It is both a logical and a metaphysical problem: The Sophists argued that if reality or Being exists only in the Absolute, then the world of experience, or sensory world, becomes pure illusion, Not-Being. If these realms are exclusive, then discriminations of truth or falsehood are inapplicable to the realm of Not-Being since Being is truth and Not-Being is false. This conclusion makes analysis and understanding of the practical world logically impossible. Plato appealed both to common sense and to logic to dispute the Sophist's contention. His logical response, according to Jowett, was a ''theory of the nature of the negative'':

> The theory is, that ''not-being is relation.'' Not-being is the other of being, and has as many kinds as there are differences in being. This doctrine is the simple converse of the famous proposition of Spinoza, not ''*omnis determinatio est negatio*,'' but ''*omnis negatio est determinatio*'';—not, all distinction is negation, but, all negation is distinction. Not-being is the unfolding or determining of being, and is a necessary element in all other things that are. We should be careful to observe, first, that Plato does not identify being with not-being; he has no idea of progression by antagonism, or of the Hegelian vibration of moments: he would not have said with Heracleitus, ''All things are and are not, and become and become not.'' (DP, IV, 384)

If Plato does not accept the idea of flux, Jowett notes that he anticipates modern philosophers in his solution to the problems of the connection of ideas and the truth of relations: ''The Platonic unity of differences or opposites is the beginning of the modern view that all knowledge is of relations; it also anticipates the doctrine of Spinoza that all determination is negation'' (DP, IV, 402).

In Jowett's view Plato's solution to the problem of Being and Not-Being is not as satisfactory as Hegel's. So Jowett concludes his introduction with a comparison of Plato's and Hegel's idealist philosophies. Their crucial similarity lies in their theories of dialectical reasoning: Plato's account of Not-Being, or difference, provided some explanation for change. But his greatest gift to philosophy, according to Jowett, ''is the recognition of the communion of classes, which, although based by him on his account of 'not-being,' is independent of this. He clearly saw that the isolation of ideas or classes is the annihilation of reasoning'' (DP, IV, 387). If Plato's dialectical mode of reasoning led only to classifications and clarifications, it nevertheless provided a model for the dynamic and progressive dialectic of Hegel. Hegel's advance over Plato, indeed, over the tradition of empiricism, lay in his theory of dialectical process—the conflict and reconciliation of contradictions—not only in the realm of mind but, as Jowett describes it, also in nature:

> Every abstraction is at first the enemy of every other, yet they are linked together, each with all, in the chain of being. The struggle for existence is not confined to the animals, but appears in the kingdom of thought. The divisions which arise in thought between the physical and moral and between the moral and intellectual, and the like, are deepened and widened by the formal logic which elevates the

> defects of the human faculties into Laws of Thought; they become a part of the mind which makes them and is also made up of them. Such distinctions become so familiar to us that we regard the thing signified by them as absolutely fixed and defined. These are some of the illusions from which Hegel delivers us by placing us above ourselves, by teaching us to analyse the growth of "what we are pleased to call our minds," by reverting to a time when our present distinctions of thought and language had no existence. (DP, IV, 406–7)

For Jowett, Hegel is the superior philosopher because he "frees the mind from the dominion of abstract ideas" (DP, IV, 413) by insisting that absolute idea can only be revealed through physical processes. That is, Hegel concretizes the dialectic; and, as Jowett's use of evolutionary terms suggests, he, like Wallace, synthesizes Hegel and Darwin.[51]

The philosophical consequences of Hegel's and Darwin's redefinition of ontological reality cannot be overestimated. Wilde followed Jowett's reasoning about one of these implications: Fixed ideas in philosophy and fixed species in nature are denied ontological status. Obviously Darwin and his critics were aware of the radical nature of his critique. "*Nullo species nova*" was a cornerstone of both Aristotelian philosophy and Christian theology. By demonstrating the development and change of species, Darwin denied this fixed nature and so the reality of any fixed category of nature.

Similarly, Hegel's dialectical system encompassed and criticized the major antecedent theories of Western philosophy. By denying language's ability to describe the world, Hegel rejected the fixed categories of previous philosophical systems like idealism or empiricism. The categories of these philosophies, however useful for analysis of the world, were ultimately inadequate descriptions of the synthetic and developmental nature of reality. Hegel's logic enabled Jowett to expose and criticize the unstated metaphysical assumptions of "common sense" philosophies. In the introduction to "The Theaetetus" Jowett describes this critical capability (we have underlined his evolutionary terminology):

> We see again that a long period in the history of philosophy was a barren tract, not uncultivated, but unfruitful, because there was no inquiry into the relation of language and thought, and the metaphysical imagination was incapable of supplying the missing link between words and things. . . . Upon the term substance at least two celebrated theological controversies appear to hinge, which would not have existed, or at least not in their present form, if we had "interrogated" the word substance, as Plato has the notions of Unity and Being. These weeds of philosophy have struck their roots deep into the soil, and are always tending to reappear, sometimes in new-fangled forms; while similar words, such as development, evolution, law and the like, are constantly put in the place of facts, even by writers who profess to base truth entirely upon fact. In an unmetaphysical age there is probably more metaphysics in the common sense (i.e., more *a priori* assumption) than in any other, because there is more complete unconsciousness that we are resting on our own ideas, while we please ourselves with the conviction that we are resting on facts. We do not consider how much metaphysics are required to place us above metaphysics, or how difficult it is to prevent the forms of expression which are ready made for our use from outrunning actual observation and experiment. (DP, IV, 154–55)[52]

Jowett's critical use of Hegelian philosophy impressed Wilde. He refers directly to this passage in his Commonplace Book in order to bolster his conclusion: "[I]t is really not a question between emplo[y]ing metaphysics, or not, but between emplo[y]ing good or bad metaphysics" (CB, 133). Wilde quotes (and emphasizes by underlining) Jowett's claim "We do not consider how much metaphysics are required to place us above metaphysics . . ." (CB, 133). To further support his conclusion and to strengthen his intellectual synthesis, Wilde also quotes the man he describes as "the precursor of the modern phenomenistic empirical school of thought," David Hume: "We must cultivate true metaphysics with some care, in order to destroy the false and adulterated" (CB, 132).

Wilde learned from Jowett the critical function of Hegelian philosophy in revealing the weakness at the foundation of nineteenth-century materialism, namely, its denial of metaphysics: "We talk about facts when we are really resting on ideas" (CB, 133). Metaphysics was for Wilde inevitable, and the only idealist philosopher who could encompass the findings of science within an evolutionary metaphysics was Hegel. Wilde noted in the Commonplace Book Wallace's description of Hegel's philosophy as "a synthesis of science but in a higher sphere—in the ether of pure thought . . ." (CB, 197).

Symonds, Pater, and Hegelian Aesthetics

In a section of the Commonplace Book Wilde used for entries on aesthetics (CB, 153–73), he copied passages from John Addington Symonds' *Studies of the Greek Poets* (1873–76) and Pater's "Winckelmann" (*Westminster Review*, 1867; reprinted in *The Renaissance*, 1873), both of which employ historical analyses of aesthetics based on Hegel's *Aesthetik* (1835).[53] Symonds and Pater were exact contemporaries as Oxford undergraduates; both had aspirations to become members of the faculty, and both learned German and read Hegel in the original.

Pater's career and writings exerted a profound influence on Wilde; they are important in providing ideas, some of which Wilde adopted and others of which he modified or reacted against in the formation of his synthesis. In "Winckelmann" Pater summarized the Hegelian theory of aesthetics and artistic progress in history and provided—especially in the 1867 and 1873 versions—quotations from and several references to Hegel's *Aesthetik* which were gradually deleted from later editions. Wilde certainly read one of the two earlier versions.[54] He recorded several passages from Pater's essay in the Commonplace Book and used them (as well as other passages from the essay) to compose "The English Renaissance in Art." A biblical reference from "Winckelmann" is echoed in the closing couplet of Wilde's poem "Hélas." Wilde noted Hegelian ideas as well as Paterian interpretations; he also added his disagreements with or extensions of these ideas, which are mostly concerned with Hegelian aesthetics, art history, and classical literature, the subjects upon which Pater's early academic reputation was based.

At Oxford Symonds' education in the *Literae Humaniores* program (1858–62) anticipated Wilde's: he studied philosophy with Jowett and the latter's student, Thomas Hill Green. He heard George Rolleston's Spencerian lectures applying evolutionary theory to comparative anatomy and physiology, and he studied Max Müller's ideas on mythology. During his preparation for the "Greats" examination, he met weekly with Jowett and recorded in his diary the influence of Jowett's teaching of Hegel.[55] After completing his examinations, Symonds continued studying Hegel, and in 1863 he proposed to Longman's "a revised and annotated, and generally metamorphosed, version of Hegel's 'Aesthetik,'" which was never completed but came to have a shaping influence on his *Studies of the Greek Poets*.[56]

Symonds' remarks in a letter show the importance of the Oxford intellectual background on the lectures he wrote at Clifton College in 1869: "I have spent today in trying to coin a theory of Mythology for the vith [*sic*] form based upon Hegel & Max Müller—Soul & Flesh. But it is an elusive Protean subject. And I cannot grapple with it, being, if any thing, poetically & not scientifically minded" (JASL, II, 51). He continued his study of Hegel while working on the second volume of *Studies*, completing Hegel's *Aesthetik* (in German) and Wallace's edition of the *Logic* in 1875. At the time he contemplated lecturing on the *Aesthetik*: "It is an extremely interesting book & full of the most brilliant things: all the discussion of Classic Art & Sculpture seems to me luminous in the last degree. But just what a critic most wants, & what ought to form the ground question of Esthetics, Hegel hardly touches upon—the principle of beauty" (JASL, II, 390). In the 1880s, still struggling toward a synthesis of Darwinism and Hegelian philosophy, Symonds wrote: "If there is truth in the Hegelian Darwinian theory of Evolution, we have no room left for caprice, for Carlylean Heroes. The great difficulty for the future will be to evaluate the forces that have to be equated—political & intellectual, climatic & temperamental, all controlled by some superior or immanent Geist of wh [*sic*] the nation and the individual are but exponents. Perhaps we shall get to do even without the Geist! But if there is no caprice, & yet a certain fixed order, we must hypot[h]esize a Geist" (JASL, II, 964).

In the 1870s and 1880s Wilde read and reviewed several of Symonds' books, beginning in 1876 with an unpublished piece on *Studies of the Greek Poets*, Vol. II.[57] In volume II Wilde noted Symonds' synthesis of Hegel's *Aesthetik* and the Darwinian and Spencerian view of evolution and heredity to explain the importance of Hellenic culture in the formation of nineteenth-century thought.[58] Wilde also cited Symonds' biography of Shelley in the College Notebook (CN, 22) and reviewed his biographies of Ben Jonson and Sir Philip Sidney, as well as the last two volumes of *The Renaissance in Italy*. Wilde sent Symonds a copy of his *Poems* (1881) and mentioned him in "The Truth of Masks," "Pen, Pencil, and Poison," and "The Portrait of Mr. W. H." Symonds admired Wilde's poetry and corresponded with him about "Mr. W. H." (which Symonds had mentioned in *Essays Speculative and Suggestive* [JASL, II, 686–87, 728–29]). When *The Picture of Dorian Gray* appeared, Symonds found it "artistically and psychologically interesting," but he disapproved of it because he thought it "unwholesome in tone" (JASL, III, 477). Wilde obviously found Symonds a man of similar interests as well as a model poet and critic who created his own synthesis of evolutionary science and idealist philosophy. In his review of the last two volumes of *The Renaissance in Italy* Wilde called him "one of the remarkable men of letters in the nineteenth century."[59]

We will supplement the Commonplace Book's Hegelian entries from Pater and Symonds with a brief description of Hegel's theory and history of aesthetics, since they form an important part of Wilde's critical position. Like the rest of his philosophy, Hegel's aesthetics were informed by idealist assumptions, dialectical logic, and a progressive and evolutionary historical perspective. In the *Aesthetik* he broke with the dominant classical and neoclassical tradition of Western aesthetic theory in two important ways. First, he argued that beauty was a product of mind and was not created by imitating nature in either its particular or typical manifestations. He believed this mimetic theory developed because an observer's contemplation, animated by love, projects beauty inherent in the mind onto the sensuous world of nature.[60] He also developed a culturally relative analysis of the history of art which defined it as a progressive (and therefore increasingly self-conscious) realization of the spirit, or ideal, in sensuous form. Wilde accepted Hegel's first innovation unhesitatingly and, with some significant modifications, also adopted the second.

Wilde adapted other Hegelian ideas. One influence comes from Hegel's discussion of the methodology of the *Aesthetik.* Previously there were two approaches to studying art: the empirical, historical, and scholarly approach, whose function was to "enrich our aesthetic appreciation of particular works of art," and the theoretical approach, which attempted to define the nature of beauty abstractly. Hegel's methodology begins in theory, but "our conception must from the beginning reconcile the two approaches we have mentioned, combining metaphysical universality with what is genuinely particular" (A, 3). In methodology, as elsewhere, Hegel's ideal is both concrete and universal.[61]

Using this method, Hegel defined art as one of the three forms of the "finite mind's awareness and apprehension of the absolute," the other two being religion and philosophy. Art is distinguished from them because it is "an immediate and for that reason *sensuous* knowing, a knowing in the configuration of the sensuous and objective itself, in which the absolute is brought to our contemplation and feeling." Religion, he says, is imaginative or "pictorial" thinking, while philosophy is "the *free* thinking of absolute mind," the most self-conscious and therefore the most advanced of the *forms* in which the absolute is brought to consciousness [emphasis in original] (A, 7). Art, then, is distinguished from religion and philosophy by its formal qualities, its way of apprehending the spirit. The artist's created work, a concrete universal, realizes the spirit in a sensuous form, which then addresses itself, through its beauty, to the spirit of the spectator.

Art has social functions, but they can only be realized through its unique formal qualities. Since "didactic utility" is discursive rather than sensuous presentation, Hegel finds it inappropriate for art, but he defines appropriate cognitive and affective functions. Art, he says, "draw[s] the human heart through the whole significance of life by means of external representations of its innermost movements," and it "purge[s] the fierceness of our impulses, passions and desires—which is to say . . . [it] is somehow nevertheless able to deliver man from the power of sensuousness" through the senses (A, 5).[62]

Hegel's definition of beauty as the sensuous embodiment of mind was, like Ruskin's, rooted in Aristotle's description of contemplation as spiritual seeing.

Great art required contemplation from both the artist and the audience. Passive observation yielded only awareness of the sensuous world, while contemplation perceived the spirit in and through the sensuous object: "[A] work of art is not naively self-centered; it is instead essentially a question, an address to the responding human soul, an appeal to affections and to mind" (A, 9). Unlike Ruskin, Hegel believed that the nature of spiritual seeing differed from one historical period to another. So, for instance, in the romantic age the spirit is manifest as a love whose essence

> consists in surrendering one's consciousness of self, in forgetting one's self in another, while yet, in that very surrender and oblivion, coming to true possession of oneself for the first time. This differs from the mediation of spirit with its other that characterizes the classical ideal, for there the spirit's other is its own bodily organism, whereas in love—which is the ideal of romantic art—spirit's other is, as we said, another spiritual consciousness, another self or person. Spirit ideally fused with body is the principle of the perfection of *plastic* beauty achieved by Greek art. In romantic art, that kind of beauty must be virtually forsaken. The fusion of spirit with spirit in love is a very different kind of principle [emphasis in original]. (A, 42)

Hegel's division of the arts into three historical stages provided Wilde with a dialectical analysis and an interpretive vocabulary which he used in his notebooks and criticism. Hegel's first stage, the symbolic, existed in Egyptian culture. He analyzed the sphinx as the self-conscious expression of the period—the symbol of symbolism—which represented the human spirit struggling to separate itself from matter and the world of nature. Symbolic art, according to Hegel, was imperfect (and knew itself to be so) in its inability to capture the spirit in sensuous form. Classicism, Hegel's second stage, is represented by Greek sculpture of the human form, in which the spirit found its perfect sensuous embodiment. This perfection realized only the individualistic and corporeal aspect of spirit and lacked an awareness of the inner world of subjective consciousness. The blank eyes of the statues demonstrate this lack, Hegel reminds us, since they are the windows of the soul. This negation in the eyes suggested the third stage of artistic development: The "spirit, no longer finding full satisfaction in the plastic beauty of the Greek gods" (A, 34), was represented by romantic (i.e., medieval and later) art as the Christian aspiration of the soul to escape the body and achieve immortality. Christ was the self-conscious representative of romantic art, the sensuous embodiment of the spirit's desire to transcend its sensuous self. The characteristic attitude of this art is love, the sacrifice of the individual self to a spiritual ideal.

Hegel recognized the contradictory character of romantic art: It represented the spirit's desire to escape the senses in a form (art) whose end is sensuous beauty. For this reason he admitted that classical sculpture was the perfect embodiment of art as a sensuous creation. He believed that religion more appropriately expressed the romantic spirit's desire to escape the sensuous world. Yet he still asserted that romantic art represented a progression in, and as, art (A, 42). Here an obvious tension exists between the cognitive function of art, namely, to show the historical evolution of spirit, and its essential formal characteristic, to represent it in sensuous form. Wilde, we believe, saw this tension and offered a historical analysis that differed from Hegel's.

Wilde used the aesthetics section of the Commonplace Book to compose his 1882 lectures on the decorative arts movement. His entries are evidence of his adaptation of various sources. Wilde uses comparative analysis, taking from Pater and Symonds the Hegelian historical progression and hierarchy of the arts (CB, 163). He distinguishes between the sculptural externality of Greek literature and art and the passionate expressiveness in medieval, renaissance, and modern forms, an expressiveness which reflects the increased attention given in later artistic forms to spiritual and psychological passions. For example, Wilde noted Symonds' comparison of Aristophanes' and Shelley's poetic depiction of clouds (CB, 155), and the differences between Euripides' and Marlowe's concreteness of characterization (CB, 157).

Wilde is most concerned with Hegel's definition of a particular, but not peculiar, Greek style, "its plastic character—it's Tendency To [*sic*] conceive ideas always under the form of images, and to give a sensible rendering to objects most removed from sense" (CB, 153). He sees it not only in Greek poetry and drama but in philosophy, sculpture, and even in the sculptor's habit of mind. He is interested in how plasticity, as the dominant ideal, shapes the mind and is embodied in various forms of expression. Greek idylls, he says, following Symonds, "are in literature what the vases are in art" (CB, 158). He also notes that in Greek drama a character is "viewed externality" [*sic*] (i.e., developed "by his relations to others not by his soliloquies"), and this externality finds its analogous expression in Greek sculpture, which "does not obtrude its muscular structure" (CB, 159).

Wilde sees, especially in drama, characteristics of Greek art which clearly anticipate their modern counterparts. For Hegel drama supplied the best transitional form because it embodied other forms of art, and for this reason he called it "the most perfect totality of content and form . . . the highest stage of poetry and of art generally."[63] Wilde follows Pater's use of Hegel in this regard by noting that "the bridge between Greek and modern life may be found in the conflict of Greek tragedy" (CB, 165). He then comments, on the following page, "of the quality and motive of Tragedy as opposed to sculpture the Greeks were keenly conscious—the calm Gods of the Parthenon looked down impassively on the passion of the Dionysiac Theatre" (CB, 167).

For Hegel drama begins in Greece and reaches its highest expression of "polished plasticity of character and action" (LFA, II, 1228) in the plays of Aeschylus and Sophocles. Wilde follows Hegel's general theory by seeing in the stage history of Euripides' *Bacchae* an important transition in the history of the artistic spirit, namely, the replacement of Hellenism by Christianity: "[W]hen the actor substituted the head of Crassus for that of Pentheus in the tragedy of the Bacchae, the material triumph of Orientalism and the sovereignty of the Parthians began: and transformed into the Christis patiens this play formed another link between East and West—and symbolized another triumph of Orientalism the Religion of Christ" (CB, 131). While Hegel thought Euripides' subjectivity was atypical of the Greek spirit (LFA, II, 1215, 1228), Wilde honored him as "the great humanist of Hellas, the cor cordium of antiquity'' (CB, 130), an influential and eloquent figure for many modern readers, poets, and translators.

By revaluing Euripides' centrality, Wilde saw more continuity and similarity between the Greek and modern spirits than Hegel did. Hegel criticized Euripides for taking a new direction in tragedy and expressing pity and emotion (LFA, II, 1215). Wilde differed from Hegel and many other nineteenth-century critics in his admiration for Euripides, who, he said, ''was criticised by the conservatives of his own day much as Swinburne is by the Philistines of ours'' (CB, 23). Wilde found Euripides' subjectivity and ability to express passion especially appealing: ''When we criticise Euripides,'' he noted, ''we must remember that it is not the highest works of the imagination which are the most affecting'' (CB, 56). Wilde's many references to Euripides in the Commonplace Book demonstrate that he regarded the playwright as expressing *both* the characteristic plasticity of the Greek spirit and the passionate emotion of the modern spirit, the ''old aesthetic ideal'' (CB, 159) and the new.

Wilde found other continuities between ancient Greek and modern cultures. His notes (and his visits to archeological sites in Greece) suggest that he may have had a more concrete and detailed knowledge of Greek art and sculpture than Hegel (who got most of his ideas from Winckelmann), and a more complex view of its development (real or imagined) than either of them. Whereas they contrasted the serenity and calm of classicism with the passion and pain of romantic art, Wilde noted the vague spiritualism and ''morbid analyzing faculty'' of René, Werther, and Faust in Euripides, and passion in the Furies of his *Orestes* (CB, 154).[64] Where Hegel identified the blank eyes of Greek statues as a sign of the absence of the soul, Wilde saw in the eyes of the river nymph of Scopes an analogue for the romantic spirit, which finds pleasure in the ''vague and misty splendour of Turner'' (CB, 157). In the same entry he notes (contra Hegel) that the Greeks depicted and enjoyed natural landscape as modern romantics have. Conversely, he identified representatives of the Greek spirit in modern art: He calls Dante, Dürer, Keats, and Blake ''the best representatives of the Greek spirit'' and sees in the Pre-Raphaelites ''frankness of detail,'' a typically Greek characteristic (CB, 156, 154). In his historical analyses Wilde is typically, almost mechanically, analytic-synthetic. He makes distinctions among historical periods, accepting each as a unified whole, but he also finds similarities among different periods and differences (or conflicts) within each. However, by stressing the continuities between classic and modern cultures, he departs from an orthodox Hegelian history of art.

Herbert Spencer and the Metaphysics of Evolution

Wilde also found in the more empirically oriented work of Herbert Spencer an evolutionary philosophy which recognized the necessity for metaphysical first principles, and thus rested on ideas rather than facts. Spencer provided biological and sociological explanations for the development of mind and the influence of the ideal in history. Although a few scholars think of Spencer as a serious philosopher, many modern accounts of his theories are so indifferent or derogatory that it is necessary to reassert the importance of his enormous influence among nineteenth-century intellectuals. His significance for Wilde appears in the entry, "English Thought from Bacon" which describes Spencer and Kant as the philosophers who carry on the "line of dialectic" by returning innate ideas to the mind (CB, 59–61).

The entry is Wilde's brief Hegelian interpretation of the development of epistemology since the seventeenth century, when inductive, democratic, and experimental thought reacted in pendulumlike fashion to the deductive, conservative philosophy of innate ideas. Hume provided a skeptical, empirical solution to the problem of how we know by claiming that, apart from sensations, mind and matter were both fictions. Kant and Spencer answered Hume's denial of innate ideas: Kant restored them to the mind transcendentally by means of the categories of the understanding; Spencer did so biologically through the "hereditary transmission of concepts." Wilde concludes with a dismissive remark about the philosopher of individualistic empiricism: "To these notions of the pendulum of thought and developement of the dialectical idea Mill cannot be said to have contributed anything" (CB, 61).

For Wilde, Spencer, not Mill, was the important thinker because his "generalistic empiricism" justified idealist conclusions by inductive means, and so represented a synthesis of the contending positions.[65] Wilde quoted from Spencer's *First Principles* in both notebooks. He seized upon Spencer's idea of the unknown as a recognition of the philosophical limitations of empiricism and materialism and as a justification for a philosophy which conceives an ultimate reality transcending human abilities to characterize it. As Wilde noted in the Commonplace Book, such a philosophical position "does not assert that the absolute is nonexistent only that it is inaccessible: from the topmost wave of empiricism we have descried the far-off shore but we have neither rudder or compass to enable us to cross the untried waters which flow between" (CB, 213). This position allowed Spencer to adopt the vocabulary and assumptions of modern science without acknowledging their ontological reality. In a College Notebook entry later incorporated in the "Rise," Wilde accepted Spencer's assertion of "Three primary truths," based on the concepts of matter, motion, and force, and adopted Spencer's argument that "Matter[,] motion and force are merely symbols of the unknown cause or reality which is coextensive with all the phenomena of the world . . . ultimately all accountable and natural facts are unaccountable and unnatural" (CN, 36–37; CW, 1124, 1144).[66]

Spencer's philosophy not only reconciled scientific assumptions with a belief in the ideal; his evolutionary theory provided Wilde with a biological explanation for the existence of innate ideas in the mind: the evolutionary mechanism of the "hereditary transmission of concepts" (CB, 61). Wilde refers to Spencer's theory of the inheritance of acquired characteristics,[67] on which, Spencer acknowledged, his entire evolutionary scheme depended. He believed that adaptive learned behavior became habitual and was passed on biologically through inheritance, and that evolutionary adaptation in human beings occurred through this process.[68] Obviously, such an evolutionary mechanism gives the mind a crucial role otherwise denied it by natural selection alone since, in Spencer's view, the mind and the adaptive strategies it develops transform matter and so are physically realized in history. In the absence of modern scientific knowledge about the mechanisms of heredity, idealists, optimists, creative evolutionists, and social reformers found the theory plausible and attractive.

Wilde referred to Spencer's theory when he noted under the rubric "Heredity," that "Spencer remarks that psychical environment is as important as physical: we cannot lay the ghosts of dead creeds" (CB, 183). Here Wilde recognized that regressive as well as progressive adaptions could be perpetuated by this process. Spencer worked out the historical and developmental significance of the inheritance of acquired characteristics in *The Study of Sociology* (1873). Basically, he "biologized" culture. Even nations, he believed, "have acquired undeniable contrasts of nature, physical and psychical, which can be ascribed to nothing but the continuous effects of circumstances, material, moral, social, on the activities and therefore on the constitution."[69]

Evidence both within the notebooks and elsewhere confirms that Wilde adopted Spencer's methods for the study of society and his conclusions about it. The Halifax, Nova Scotia *Morning Herald* (10 October 1882) reported that during his visit Wilde spoke "of sociology, of Herbert Spencer, whom he had read and admired greatly. He found nothing in his work or in any other work on evolution which differed from Plato and Aristotle."[70] The Commonplace Book also contains several entries based on *The Study of Sociology*. Wilde was interested in Spencer's "generalistic empiricism" as well as his explanation of the necessary development of individual and social organisms from simple to complex, from the chaotic and unformed to the highly differentiated and interdependent. Wilde recorded Spencer's analogy between the individual and the social organism, and since he accepted the theory of the inheritance of acquired characteristics, he followed Spencer in postulating that "the science of society . . . rests on the science of life: sociology on Biology" (CB, 7).

Wilde also accepted Spencer's assumption that the study of society could be truly scientific. He noted Spencer's various objections to the scientific study of social aggregates—assumptions of free will, "the Great Man theory of Life," and "extra-natural interference"—and agreed with the generalistic empiricist assumptions that society is an interdependent organism in which causation can be observed and about which some predictions can be made. By adopting this position, Wilde rejected both supernatural interventions and liberal atomistic assumptions about the nature of society in the "individualistic empiricism" of Mill and other utilitarians (e.g., that since the individual was the basic sociological unit, all social aggregates were unscientific fictions).

W. K. Clifford and Moral Chemistry

Spencer's account of the mind's role in the process of adaptation gave Wilde the keystone for his synthesis of evolutionary theory and the essentialist philosophies of Plato, Aristotle, and Hegel. Spencer's theory of evolutionary change, which described an organic purposiveness in the individual and society, and the possibility of perfection in both, could be interpreted as a modern extension of Aristotle's evolutionary theory compatible with Hegel's developmental philosophy. Understandably, Wilde was interested in a young scientist who also accepted Spencer's theories of the inheritance of acquired characteristics and of societies as organisms.[71] William Kingdon Clifford, whose brief life ended about the time Wilde was compiling his notebooks, wrote brilliant essays on the evolutionary development of mind, the philosophical limitations of science, and the scientific basis of morals. Wilde read extensively in Clifford's *Lectures and Essays* (1879) and noted further evidence to counter the prevailing individualistic assumption which dominated liberal political, philosophical, and sociological theory, namely, that a person's identity could be established in opposition to society or apart from it. Clifford's theory of the moral sense also attracted Wilde because it provided an evolutionary basis for intuitionist ethics.

Wilde drew from Clifford's "On the Scientific Basis of Morals" (1875) an evolutionary analysis of the development of individual conscience through tribal survival. Basing his argument on some remarks in Darwin's *Descent of Man*, Clifford theorizes that an individual's moral sense, by which he means "the Categorical Imperative," piety, or conscience, evolved because it secured the preservation of the tribe: "In the first place, ethical maxims are learned by the tribe and not by the individual. Those tribes have on the whole survived in which conscience approved such actions as tended to the improvement of men's characters as citizens and therefore to the survival of the tribe. Hence it is that the moral sense of the individual, though founded on the experience of the tribe, is purely intuitive; conscience gives no reasons" (L & E, II, 119). These intuitions or social instincts which Clifford called the "tribal self" were "something in the mind of each individual man which binds together his gregarious instincts" (L & E, II, 112). Clifford's theory, namely, that the survival of the tribe occasionally required individuals instinctively to sacrifice personal interests in favor of the group, explains why people perform "other-regarding acts," a problem J. S. Mill could never fully account for on the basis of self-interest and personal pleasure.

In a section of the Commonplace Book entitled "Moral chemistry," Wilde synthesizes information from two of Clifford's essays. He notes, from "On the Scientific Basis of Morals," that individualism and its outgrowths, "private property and a private conscience," are not fundamental to human ethical nature: "[I]t is the Tribal self wh. is the first mainspring of action, and canon of right and wrong: a savage is not only hurt when a man treads on his own foot, but when the foot of the tribe is trodden on" (CB, 119). Following Clifford, Wilde identifies the categorical imperative as the modern manifestation of the tribal self and considers the "preservation of the Tribe" rather than the "greatest happiness of the greatest number" as the ethical principle sanctioned by evolution (CB, 121). Wilde used Clifford's explanation and justification of intuitive ethics to develop further his reconciliation of evolutionary theory and philosophical idealism. Since evolutionary and ethnological science, as Darwin, Spencer, and Clifford understood it, emphasized the inheritance of survival characteristics such as moral behavior, Wilde could accept the idea that the social and cooperative, as opposed to the individualistic, aspects of human behavior were not only necessary to survival but *a priori,* or instinctive. Such a theory, of course, ran counter to the assumptions of mid-Victorian liberalism.

Clifford also contributed to contemporary revisions of the philosophy of science.[72] In "The Philosophy of the Pure Sciences," originally presented as lectures in 1873, he rejected the naive positivism of earlier thinkers and insisted upon a pragmatic-economical view: Science was a practice with no metaphysical basis. Positivism, then, neither denied nor affirmed metaphysics, and scientists could and did differ about metaphysical matters.[73] Clifford called himself an idealist and his theory appears in Frederick Pollock's biographical introduction to *Lectures and Essays*:

> Briefly put, the conception is that mind is the one ultimate reality; not mind as we know it in the complex forms of conscious feeling and thought, but the simpler elements out of which thought and feeling are built up. The hypothetical ultimate element of mind, or atom of *mind-stuff*, precisely corresponds to the hypothetical atom of matter, being the ultimate fact of which the material atom is the phenomenon. Matter and the sensible universe are the relations between particular organisms, that is, mind organized into consciousness, and the rest of the world. This leads to results which would in a loose and popular sense be called materialist. But the theory must, as a metaphysical theory, be reckoned on the idealist side. To speak technically, it is an idealist monism. Indeed it is a very subtle form of idealism, and by no means easy of apprehension at first sight [emphasis in original].[74]

Clifford also knew the German idealist theories of Schelling and Hegel. In "The Philosophy of the Pure Sciences" he explained in a materialist way their conceptions of consciousness (L & E, I, 293). In two different entries (CB, 125, 137), Wilde copies from or paraphrases Clifford's comments about the problem of consciousness, and the scientific and philosophical understanding of it:

> It seems to me that the Kantian dilemma about universal propositions is just as valid now . . . as it was in his time. How am I to know that the angles of a triangle are exactly equal to two right angles under all possible circumstances; not only in those regions of space where the solar system has been, but everywhere else? The accumulated experience of all my ancestors for a hundred and fifty million years is no more competent to tell me *that* than my own experience of the last five minutes. Either I have some source of knowledge other than experience, and I must admit the existence of *à priori* [*sic*] truths, independent of experience; or I cannot know that any universal statement is true. Now, the doctrine of evolution itself forbids me to admit any transcendental source of knowledge; so that I am driven to conclude in regard to every apparently universal statement, either that it is not really universal, but a particular statement about my nervous system, about my apparatus of thought; or that I do not know that it is true. (L & E, I, 281)

Wilde follows this quotation by adding, "Now here he is at one with the idealists" (CB, 125). Clifford had, in fact, effected a physiological explanation of *a priori* and idealist metaphysics. As Wilde says elsewhere, "The experience of the race having been substituted for the experience of the individual, necessary truths are admitted to be a-priori to the individual, though a-posteriori to the race . . .

a universal statement is really a particular statement about the nervous apparatus of thought'' (CB, 137).[75] There can be no doubt that Wilde understood how Clifford's work was thus quite compatible with philosophical idealism as it was developing in the thought of the Oxford Hegelians.

Thomas H. Huxley and the Idealism of a Materialist

Clifford described the implications for the philosophy of science of his physiological explanation of necessary truths, and Wilde, in his entry ''Necessary Truths: unwarrantable psychological facts'' (CB, 137–41), used Clifford's reasons to counter T. H. Huxley's explanations for such truths. Wilde knew Huxley's explanations because he had just read *Hume* (1879), ''On the Physical Basis of Life'' (1868), and other essays as preparation for writing ''The Rise of Historical Criticism.'' The *Hume* entries that appear in both notebooks are mainly concerned with evolutionary theory, psychology, and the philosophy of science. They also contain ideas which support Wilde's synthesis, including Huxley's confirmation of the theory of social organism (CN, 53) and an extended description of ''Matter and Mind'' amenable to an idealist metaphysics (CN, 57–63).

The latter entry offers particularly revealing evidence of Wilde's interest in and use of the writings of an important scientist who practiced materialism in science but preached agnosticism in philosophy. In discussing the problem of matter and mind, Huxley defines the materialist position of most working scientists when he states that ''the roots of psychology lie in the physiology of the nervous system.''[76] However, in his later chapter on ''The Soul: The Doctrine of Immortality'' he describes four other explanations for the relation of mind and matter, only to conclude that Hume's agnostic position is correct: ''[W]e do not, and cannot, know anything about the 'substance' either of the thinking thing or of the extended thing'' (*Hume*, p. 164).

Wilde's long ''Matter and Mind'' entry (CN, 57–63) incorporates Huxley's description of the various solutions to the problem—and objections to them—before, during, and after Hume. Wilde does not credit Huxley or Hume directly but refers to ''modern scientists'' when he summarizes or quotes their positions. He also quotes John Tyndall, another of the ''modern scientists,'' on the problem and paraphrases Hume's and Tyndall's solutions. Read in the intellectual context we have established, Wilde's entry is a statement by scientists about the limitations of scientific knowledge which leaves open the possibility for an idealist's assertion of the spiritual reality of mind. Furthermore, the three ironic comments and questions Wilde writes beside this entry suggest that he was not satisfied with positivist scientific explanations. While positivists might describe physical antecedents or accompanying characteristics, they could not fully account for the meaning of human experience (CN, 56). His question about sorrow and seasickness presumes that a meaning exists beyond involuntary physical responses. The second question, about the prophetic voice crying in the wilderness, suggests that no purely physical analysis of the cause of speech can account for its essential purpose (cf. our earlier discussion of Müller's theory of language). The third comment directly states the limited value of physical analysis and so implies a question: What do we really learn or say about anger when we describe its bodily manifestations? All three comments imply that an adequate analysis of human behavior requires, in addition to the physical, a mental or spiritual cause.

The last entry in the Commonplace Book summarizes Wilde's difficulties with the position of "modern science" on matter and mind. Science cannot explain "the consciousness of a persistent ego underlying all sensations and different from them." That is, science has found parallel material changes inside and outside organisms, but "this does not solve the problem of consciousness" (CB, 217).[77] Another example of Wilde's interest in the limitations of science appears in his entry on "Miracles," which, it was popularly believed, had been disproved by science. Wilde first noted Huxley's refutation of Hume's argument from probability and then his denial that a miracle could violate a law of nature as examples of a scientist's reasons for admitting the possibility of miracles.

Wilde respected the truth and authority of modern science and would not assert an idealist philosophy at odds with contemporary scientific knowledge. So, for instance, he remarks, "There is no need to assume an immaterial liver behind the hepatic structure" (CB, 149). However, he felt that idealists should rush in where materialists feared to tread when the constraints of materialists' assumptions prevented them from explaining certain human psychological experiences. So, Wilde wrote, "Modern Positivists are as men who while they deny the existence of the sun yet worship the sunlight on the Earth, who acknowledge that the fruit is sweet to eat, and the flower goodly for sight yet insist that the root is rotten, and the soil barren" (CB, 89). He developed his synthesis to provide a sun, a soil, and roots adequate to explain the bounty they produced.

Wilde's Synthesis

Having examined philosophical and scientific writings which influenced Wilde, we can reconstruct from these two notebooks the synthesis he was building. The work of his parents, their friends in Dublin, and of J. P. Mahaffy at Trinity introduced Wilde to subjects he continued to study, the newly developing sciences of society and the history of ancient and modern philosophy. If he abandoned some early theories in part or whole, he nevertheless took from them an expectation that any account of human experience required systematic and synthetic explanatory power which both these traditions assumed. This theoretical expectation probably stimulated Wilde's interest in the theories of Max Müller and John Ruskin: In different ways, both men attempted to fit into their idiosyncratic idealist beliefs certain findings of modern science about historical philology, myth, and the racial inheritance of morality, imagination, and the art-gift.

At Oxford Wilde also found in William Wallace's and Benjamin Jowett's explanations of Hegelian theory a variety of idealism better fitted than Müller's and Ruskin's to incorporate the materialist assumptions and findings of science, and especially evolutionary theory. Furthermore, Wilde adapted Hegel's aesthetic theory and a modified version of his history of art, which he learned either firsthand or indirectly from Walter Pater or J. A. Symonds. Wilde tested this synthesis primarily by investigating the philosophical writings of critical materialists like Huxley, Tyndall, and Spencer. He found theoretical and factual support in the scientists' own awareness of the limitations of their knowledge and theories. He also found in Spencer's evolutionary theory, and especially in W. K. Clifford's development of it, a scientific explanation for the development of the idea in history, namely, the inheritance of acquired characteristics. Wilde's position is very close to Clifford's, but he adds to Clifford's incipient reconciliation of idealism and positivism the concept of absolute idea slowly evolving toward perfection in history. The inheritance of acquired characteristics provides the mechanism for this progressive development, and the assumption of a unity between mind and matter, ego and non-ego, "mind-stuff" and the organic molecules of the brain provides the physical and spiritual location in individuals for absolute mind.

In accepting the Spencer-Clifford line of evolutionary theory, Wilde places himself in an essentialist tradition of biological explanation (organic purposiveness) which begins with Aristotle. In attempting to develop and extend this tradition in the notebooks, Wilde enacts the Hegelian theory he accepts. This practice, we contend, prepares for and is analogous to the Hegelian critical position informing "The Rise of Historical Criticism." There Wilde employs a method of historical analysis to describe and criticize its own origins and development in the writings of the classical historians and philosophers. In both history and critical theory Wilde assumes an idealist interpretation of evolution (the heart of his synthesis), the parallel development of matter and mind toward perfection:

> Just as the human embryo passes through all the stages of evolution from the lowest to the highest organism during the progress of its perfection, so the human mind must pass through all the stages which have been already taken in the progress of the intellectual world: neither in the world of thought or in that of matter is the past ever annihilated: progress in both must be made by slowly graduated stages from simple sensations, and formless protoplasms, to the highest differentiated organism and the purest abstractions of thought. (CB, 83)

3

The Text as Context

In this chapter we shall trace the influence of Wilde's synthesis on his published criticism and fiction. In describing the synthesis and its background, we have cited selectively from the notebooks. Many other entries have a complementary relation to the synthesis: These Wilde used to write the "Rise" and to prepare himself as a literary humanist, which, for many Victorians, meant a classicist. The notebooks contain information about and meditations upon Hellenic culture which Wilde used to develop his cultural ideal, the "New Hellenism," in "The Rise of Historical Criticism," "The English Renaissance in Art" and his other lectures in America, *Intentions,* "The Soul of Man Under Socialism," "The Portrait of Mr. W. H.," and *The Picture of Dorian Gray.*

"Hélas" and the New Hellenism

We concur with Richard Ellmann that the first poem in Wilde's collection of 1881, "Hélas," provides an excellent introduction to Wilde's writing because it concerns the relation of Hellenism to modern culture. We disagree with Ellmann's reading of this sonnet as Wilde's halfhearted rejection of classicism in favor of Pater's worldview and his aesthetic impressionism.[78] "Hélas" may be Wilde's French sigh of modernist deliquescence, but it is also a suggestive pun on "Hellas," and we believe that in this poem Wilde addresses himself to the problematic relationship between the Hellenic past (which he identifies with spirit, maturity, and order) and the modern age (which he identifies with romanticism, passion, and immaturity). We agree with Ellmann that "Wilde considered Hellenism to be the more basic side of his nature, overlaid, but only as a palimpsest conceals the original, by a more modern mode."[79] However, we believe that Ellmann has systematically underestimated Wilde's commitment to Hellenism. In the poem Wilde sees his life as a "twice-written scroll," and in a series of contrasting images he questions whether his fleeting experience of boyish passion and romance will cause him to "lose a soul's inheritance":

> To drift with every passion till my soul
> Is a stringed lute on which all winds can play,
> Is it for this that I have given away
> Mine ancient wisdom and austere control?
> Methinks my life is a twice-written scroll
> Scrawled over on some boyish holiday
> With idle songs for pipe and virelay
> Which do but mar the secret of the whole.
> Surely there was a time I might have trod
> The sunlit heights, and from life's dissonance
> Struck one clear chord to reach the ears of God:
> Is that time dead? lo! with a little rod
> I did but touch the honey of romance—
> And must I lose a soul's inheritance? (CW, 709)

The notebooks and the "Rise" provide a context that confirms what the images and rhetoric of the poem also strongly suggest: the importance of the Hellenic past and the relative insignificance (even decadence) of modern experience. "Ancient wisdom," "austere control," and even "the secret of the whole" were what Wilde had striven to realize at Oxford and to reveal in the "Rise." According to his racialist theory of culture, they literally constitute the "soul's inheritance." The contrasting images (drifting, the Aeolian harp, idle songs and holiday) all suggest a lack of control, a giving in to emotion and immediate gratification. As Wilde writes, they "mar" that inheritance which would allow him to "reach the ears of God." Rather than celebrating this modernist drift, the poem laments it (Hélas) and the possible loss of this inherited soul (Hellas).

An allusion in the poem suggests that Wilde may have intended it as a response to Walter Pater's "Winckelmann" (1867). In lines twelve and thirteen Wilde paraphrases a biblical quotation that Pater used in his essay; in doing so, he points up his differences with the sage of Brasenose. Pater wondered if it was possible for modern man to regain the Greek attitude toward sensuousness:

> To the Greek this immersion in the sensuous was indifferent. Greek sensuousness, therefore, does not fever the blood; it is shameless and childlike. But Christianity, with its uncompromising idealism, discrediting the slightest touch of sense, has lighted up for the artistic life, with its inevitable sensuousness, a background of fame. "I did but taste a little honey with the end of the rod that was in mine hand, and lo, I must die." It is hard to pursue that life without something of conscious disavowal of a spiritual world. . . .[80]

Pater's quotation from 1 Sam. 14 suggests that his rejection of God for sensuousness is a good thing. Jonathan, who speaks this line, has unknowingly broken the commandment of his father, Saul, by eating honey, and has broken God's law by encouraging the Israelites to eat bloodied meat. However, the result has been enlightenment for Jonathan and a great victory for the Jews over the Philistines. Because Jonathan is not punished, the story sanctions the selective breaking of God's law, and Pater uses it to justify the defeat of modern Philistinism through the pursuit of the sensuous pleasures of classical statuary.

In the poem Wilde paraphrases the same quotation as a question but changes its context and meaning. Wilde sees himself as Hellenic and identifies his heritage not with sensuality but with the wisdom and austere control which will allow him to rise to spiritual heights and so "to make music that will reach the ears of God." Hellenic art, for him, is the vehicle of the spirit. In this context the biblical paraphrase implies an answer to the question of the last two lines. Like Jonathan, he will not lose his soul's inheritance. His Hellenic past, instinct with spirit, is not threatened by "the honey of romance" and could be revived in the modern world. With it would come the wisdom, control, and idealism which create and are embodied in great art.

"Hélas" prefaces Wilde's first book. It announces his constant theme, the creative and critical efforts to establish a New Hellenism which represents an antithesis to the fallen world of the present, at once a dialectical return to and advance forward from the golden age of Greece. Frank M. Turner has described three Victorian interpretations of Hellenic civilization as Arnoldian, aesthetic, and evolutionary.[81] Wilde was among the "evolutionary" interpreters of that culture. Whereas the older Arnoldian and aesthetic interpretations had focused primarily on Greek literary and artistic culture as a repository and model of timeless values, evolutionary interpretations, based upon mid–nineteenth-century archeological discoveries and later anthropological theories, emphasized its origins and the dynamics of its development.[82]

The "Rise," the Notebooks, and Historical Criticism

We previously outlined the dialectical structure and content of "The Rise of Historical Criticism" to provide a context for explaining the notebooks. Now we shall reexamine the essay, attending to topics or problems Wilde raises about historical criticism: its origin, its nature, and its relation to science, philosophy, political theory, and aesthetics. Our analysis will confirm the centrality of Wilde's synthesis and its components (mind, imagination, race, and evolution) as guiding ideas in the "Rise." Even though it was unpublished in his lifetime, the essay was his first realization in critical discourse of the synthesis recorded in the notebooks; each helps to explain the other, and together they provide a context for interpreting the meaning of his later works.

We begin, then, with origins. Wilde warns, early in the "Rise," that "*L'esprit d'un siècle ne naît pas et ne meurt pas à jour fixe,* and the first critic is perhaps as difficult to discover as the first man" (CW, 1106); but he locates the first manifestation of the spirit of historical criticism in "the Hellenic branch of the Indo-Germanic race" (CW, 1105). So, in the beginning Wilde defines criticism as a spirit; his evolutionary outlook is suggested by his interest in the origins of spirit, by his recognition of the difficulty of finding them, and by his acceptance of a theory of prehistoric racial origins. This statement and his further identification of Aryan origins for the "Greek race" (CW, 1106) place Wilde in the tradition of idealist racialists like Ruskin and Müller, who assume that racial inheritance of mind provides a satisfactory explanation for the spirit's historical continuity and power. The critical spirit first appeared as "a changed attitude of mind [rather] than as a principle of research" which "burst on the Greek race like a great flood of light about the sixth century B. C." (CW, 1106; cf. CB, 21). Wilde suggests that the critical and creative insights arising from this "speculative faculty" produced the "most truthful" kinds of early history, and that Aeschylus and Lucretius "attained to truth by a certain almost mystic power of creative imagination . . ." (CW, 1121–22). He asserts that the germ of scientific understanding "seems to have come to the Greek mind spontaneously, through a certain splendour of imagination . . . before inductive research had armed them with the instruments of verification" (CW, 1122). This germ gave rise to the "spirit of skepticism," which denied religious and miraculous explanations of nature and history and instead provided naturalistic interpretations for religious and secular historical writings.

In the College Notebook Wilde remarks, concerning the unity of creative and critical powers, that "the influence of physical science is rather the attracting influence of a new analogy than a practical disproof of any particular belief." He asserts that the "force of analogy[,] the desire to bring all ones [*sic*] thoughts into harmony, and mutual correspondance [*sic*], have led men to infer that the reign of Law . . . is also to be extended to those phenomena which seem the most remote from Law" (CN, 105). The attraction of analogy, which Wilde uses to link science with other forms of thought, has at its root the idealists' belief in the fundamentally harmonious structure and development of reality. Like the English and German Romantics, Wilde sees the imagination, the analogizing power, as the fundamental quality of mind.

This critical-creative union informs the theory and structure of Wilde's later work and explains his attempts to develop dialectically such forms as dialogues, fictionalized biographical and critical studies, the novel of ideas, and even aphorisms. A title such as "The Critic as Artist" reveals his assumption of this relationship and its continuity in his critical thought. In that essay he insists on the necessity of the critical faculty for the creative writer and of the creative aspect of critical writing. Wilde sees criticism as serving an essentially antagonistic or antithetical intellectual and social function. The spirit is realized in and through opposition to custom, habit, and stasis. Wilde adopted an appropriate rhetorical stance which characterizes his later work, namely, taking elements of mid-Victorian political, philosophical, and scientific beliefs and developing their subversive implications.

Wilde argued that this critical-creative spirit is first manifest as "the historic sense," an intuitive and racially inherited attitude of mind which first appears in written histories and only later is rationalized as the science of historical criticism, that is, a self-conscious methodology and philosophy. Wilde defines historical criticism teleologically as part of Mind, World-Spirit, or Idea manifesting itself in history—"that complex working towards freedom which may be described as the revolt against authority" (CW, 1105). He defines its characteristics, in both methodology and attitude of mind, in relation to other speculative, innovative, and progressive manifestations of that spirit: democracy, revolution, philosophy, and physical science. From them it takes its essential qualities, an "intolerance of dogmatic authority," a belief in "the alluring analogies of law and order," and a "conception of an essential unity underlying the complex manifestations of phenomena" (CW, 1106).

The essential qualities Wilde ascribed to historical criticism derive from his synthesis of science and philosophy. He used his notebooks to identify modern problems of historiography, to suggest solutions, and to quote, summarize, and paraphrase ideas and examples relevant to the essay. One such problem is law and order in historiography. In the "Rise" Wilde points to a contradiction in Western thought: "Now at the outset we must note that, while the conceptions of law and order have been universally received as the governing principles of the phenomena of nature in the sphere of physical science, yet their intrusion into the domain of history and the life of man has always been met with a strong opposition . . ." (CW, 1122). In the essay and the notebooks Wilde lists four reasons generally given for rejecting the doctrine of uniform sequence: belief in free will; in "extranatural interference" (or the "theological doctrine of irregularity"); the "Great man theory"; and the adoption of a pathological view of events (CW, 1122, 1136; CN, 3, 29–30; CB, 5). In the "Rise" Wilde rejects all four and states that "the very first requisite for any scientific conception of history is the doctrine of uniform sequence" (CW, 1122).

The notebooks show how Wilde explored the issue and related it to his position that only an inductive-deductive method is adequate for writing history. They record statements from four modern historians (H. T. Buckle, W. E. H. Lecky, J. A. Froude, Robert Flint), an anthropologist (E. B. Tylor), a sociologist (Herbert Spencer), and a Greek philosopher (Aristotle) which provide authority for Wilde's assertion in the "Rise" of the doctrine of uniform sequence (CN, 29, 3; CB, 3). These entries provide support from various disciplines and show his characteristic interest in intellectual convergences. Interestingly, in the "Rise" and the notebooks Wilde does not record the long rationales provided by these men, even though he was aware (as the notebooks and the essay show) that "uniform causation" in philosophy and history was problematic. He might have excluded them because he believed that first principles could not be proved or because his interest in the classical origins of historical criticism did not require the restatement of modern problems. We believe, however, that Wilde's selections were dictated by his Hegelian assumptions about how ideas are realized in practice.[83] In the essay he shows the "idea" of uniform sequence developing from ancient religions to the philosophy and practice of classical historians. Wilde probably believed that if modern history was to become scientific, it had to recover such a methodology, one already realized in the practice of modern science. He assumed that a true scientific method relied upon a belief, an act of faith, in the orderliness of the universe. Without such a faith or, in modern terms, without a paradigm, science is reduced to a mere collection of facts. One must assume orderliness before one can find order since, in the phenomenal experience of scientists, disorderliness can only be seen as a problem if one assumes, believes, or can imagine another, more orderly state of affairs.

In his notebooks Wilde listed examples from the history of science which illustrated how scientists who assumed the doctrine of uniform sequence solved problems by finding regularity where others found only irregularity. He noted Buckle's ironic description of the difference between natural scientists' acceptance of uniform sequence and historians' reliance on the doctrine of irregularity (CN, 29). On the next page he recorded Buckle's examples of two eighteenth-century Scottish scientists, William Hunter and William Cullen, who found that regularity existed in pathology (i.e., in the study of the abnormal) by assuming that all of nature obeyed the doctrine of uniform sequence. Buckle contended (and Wilde concurred) that a true scientific method must assume that "all science is manifestly converging towards one simple and general theory, which shall cover the whole range of material phenomena, and that, at each successive step, some irregularities are explained away . . . that life is a property of all matter, and that the classification of bodies into animate and inanimate, or into organic and inorganic, is merely a provisional arrangement. . . ."[84] Buckle cited the achievements of Hunter and Cullen in the study of animal monstrosities and the apparent irregularities in crystals as a dramatic validation of this principle (CN, 11). He explained that they had to adopt an inductive-deductive method rather than either alone since "it has been found, that the best way of arriving at a theory of disease is by beginning with the theory of health; and that the foundation of all sound pathology must be first sought in an observation, not of the abnormal, but of the normal functions of life" (HCE, I, 363). He suggested that historians of social development ought to adopt this paradigm.

In the "Rise" Wilde paraphrases Buckle's observation that Polybius, following Aristotle, "may be said to have anticipated one of the most important truths of the modern methods of investigation: I mean that principle which lays down that just as the study of physiology should precede the study of pathology, just as the laws of disease are best discovered by the phenomena presented in health, so the method of arriving at all great social and political truths is by the investigation of those cases where development has been normal, rational, and undisturbed" (CW, 1136). He also explains that Aristotle's historical method, "the true method," is "neither exclusively empirical nor exclusively speculative, but rather a union of both in the process called Analysis or the Interpretation of Facts" (CW, 1125). For Wilde this can only be so because Aristotle assumed "that nature, including the development of man, is not full of incoherent episodes like a bad tragedy, that inconsistency and anomaly are as impossible in the moral as they are in the physical world, and that where the superficial observer thinks he sees a revolution the philosophical critic discerns merely the gradual and rational evolution of the inevitable results of certain antecedents" (CW, 1125; cf. CN, 3, 29).

Although Wilde drew upon Buckle's prose to describe Aristotle's methods and assumptions, the notebooks also show that he found support for his evolutionary theory of history in other modern sources, such as the writings of E. B. Tylor, Herbert Spencer, and Robert Flint. Wilde uses Tylor and Spencer to counter the theological argument that "extra-natural interference" or "the power of free will" deny the possibility of lawfulness in history (CW, 1126, 1122; CN, 3; CB, 5). In the College Notebook (CN, 21) Wilde paraphrases Flint's position to counter another aspect of the theological theory: "The philosophy of history is not a something separate from the facts of history, but a something contained in them."[85] By synthesizing these positions with Aristotle's and Hegel's, Wilde imaginatively revises an idealist philosophy of history. History, he says, is "an organism containing the law of its own development in itself, and working out its perfection merely by the fact of being what it is" (CW, 1126).

To support his inductive-deductive methodology Wilde takes issue with historians who believed that the scientific method consisted solely in the collection of facts. As Wilde says in the "Rise" and his notebooks, one problem of modern history is the "enormous complexity of the facts" (CW, 1137). He twice notes Leibniz's remark that "the attaching too high a value to mere facts is often a sign of a want of ideas" (CN, 27, 44). In the "Rise," however, he makes it equally clear that he would not ignore or destroy facts to make the historian's task easier. Great historians like Montesquieu, he believes, are capable of detecting important relations among facts which seem insignificant to "the vulgar" (CW, 1137; CN, 65). So he discusses the relation of facts and ideas in history based on the premise that

> what is important in all great writers is not so much the results they arrive at as the methods they pursue. The increased knowledge of facts may alter any conclusion in history as in physical science, and the canons of speculative historical credibility must be acknowledged to appeal rather to that subjective attitude of mind which we call the historic sense than to any formulated objective rules. But a scientific method is a gain for all time, and the true if not the only progress of historical criticism consists in the improvement of the instruments of research. (CW, 1135)

Since Wilde believed this subjective "historic sense" was a manifestation of the "critical-creative" spirit, the proper practice of history inevitably involved the interaction within the historian of theoretical ideas and the collection of facts, both disciplined by the scientific method. The historic sense intuits the relations among facts and so generates the theory which guides factual selection; the final interpretation must be tested by its relation to facts known at the time or discovered later. Wilde offers Polybius as a model historian who adopted "neither the purely deductive nor the purely inductive mode but the union of both" (CW, 1138). The latter recognized that "a certain selection must be made, and a selection involves a theory" (CW, 1138). So Polybius, applying this method, "points out that those phenomena particularly are to be dwelt on which may serve as a *παράδειγμα* or sample, and show the character of the tendencies of the age as clearly as 'a single drop from a full cask will be enough to disclose the nature of the whole contents.' This recognition of the importance of single facts, not in themselves but because of the spirit they represent, is extremely scientific . . ." (CW, 1138; cf. CN, 77–81).

Wilde saw Polybius as "the culmination of the rational progression of Dialectic" (CW, 1133). Polybius' inductive-deductive procedure constituted in all but name a dialectical method. In consciously employing the abstract method for selection, he simultaneously assumed "the fact which is usually urged as the strongest objection to the employment of the abstract method,'' namely, ''the conception of a society as a sort of human organism whose parts are indissolubly connected'' (CW, 1138). Although Polybius recognized the practical implications of this method for historiography, it remained for Plutarch ''to complete the progress of thought'' (CW, 1143). Plutarch made the metaphysical foundations of Polybius' method explicit: ''In a word, Plutarch felt that while science brings the supernatural down to the natural, yet ultimately all that is natural is really supernatural'' (CW, 1144).

Wilde's overall strategy in the ''Rise'' (if not to turn Hegel on his head, then to turn him inside out) ends with his dialectical connection of Plutarch and Herbert Spencer. Paraphrasing Spencer as ''the last message of modern science,'' Wilde writes that '' 'when the equation of life has been reduced to its lowest terms the symbols are symbols still,' mere signs, that is, of that unknown reality which underlies all matter and all spirit. . .'' (CW, 1144; cf. CN, 37). In reaching this conclusion, Wilde incorporated the spiritual and imaginative sources for a scientific methodology of history: ''[T]he first and the last word of Greek history was Faith'' (CW, 1145). This sense of circularity, where ends and beginnings are joined, is fundamental to Hegelian analysis and is also manifest in the rhetorical and critical stance of the ''Rise.'' There, when Wilde found the roots of his own, modern historical criticism in the works of classical historians and philosophers, it was partly because he assumed that the roots existed beneath the tree. He, like the Wordsworth he wisely invented, ''found in stones the sermons he had already hidden there'' (CW, 977–78).

Dialectical Criticism After the "Rise"

Most critics have not understood the formative importance of the ''Rise'' or the notebooks in Wilde's intellectual and critical oeuvre. They have limited their attention to literary and artistic traditions such as aestheticism, symbolism, and decadence, and so have exaggerated the importance of one strand in the complex fabric of Wilde's critical position. In their search for a usable literary past, they have defined Wilde as an art-for-art's-sake critic who is a precursor of modern critical theory and a major bridge from Victorian aestheticism to modern aesthetics.[86] Here we argue that Wilde's critical position properly belongs not to the art-for-art's-sake movement but to the traditions of late Victorian cultural criticism. His dialectical method and his synthesis of idealism and science constitute the unchanging first principles of much of his published work and especially of his theories of literature, art, and criticism. In this section we will reexamine Wilde's intellectual relationship to Walter Pater and the intellectual assumptions of art-for-art's-sake criticism. We will show how Wilde developed scholarly and creative forms of dialectical criticism in his lectures in America and in *Intentions* (1891).

In histories of literary criticism, art for art's sake is usually described as a strand of Arnoldian humanism which survived the more skeptical and sophisticated philosophical challenge of late–nineteenth-century thinkers; Wilde's critical ideas have often been explained as a witty reformulation of Pater's aesthetic impressionism. As such, Wilde's theories are said to represent what remained of Arnold's ideas and values after Pater's skeptical and philosophically acute mind had tested them. As Wendell Harris and Richard Ellmann, two major proponents of this interpretation, have suggested, Arnold's stipulation that the critic "see the object as in itself it really is" was reinterpreted by Pater in the "Preface" to *The Renaissance* to mean that one should "know one's own impression as it really is." Both contend that in "The Critic as Artist" Wilde's insistence that the critic "see the object as in itself it really is not" does not differ substantially from Pater's statement.[87] This epistemological position leaves the aesthetic theorist without a philosophical basis for a mimetic theory of art and so can lead directly to impressionism, to the detachment of art from life, to aestheticism and formalism. Imprisoned in a consciousness which is robbed of mimetic ability, both artist and critic are condemned to individual creativity.[88]

This position accurately describes Pater's revision of Arnold, but Wilde, we shall argue, disagreed with Pater's conclusion and its implications for artist and critic. Pater, like his tutor, Jowett, was concerned with the relationship of idealism and Darwinian evolutionary theory. His knowledge and use of science, especially Darwinian theory, appear in his earliest essay, "Coleridge." In his introductory paragraph Pater speaks of a "law of development" in Nature which "evolves ideas, hypotheses, modes of inward life, and represses them in turn. . . ."[89] He then distinguishes "relative" modern from "absolute" ancient thought by denying the reality of "classification by 'kinds' or *genera*." He concludes: "The philosophical conception of the relative has been developed in modern times through the influence of the sciences of observation. Those sciences reveal types of life evanescing into each other by inexpressible refinements of change. Things pass into their opposites by accumulation of undefinable quantities" (H & T, 1438).

The skepticism of positive science for metaphysics also deeply influenced Pater. In the "Preface" to *The Renaissance* he asserted, "Beauty, like all other qualities presented to human experience, is relative; and the definition of it becomes unmeaning and useless in proportion to its abstractness" (TR, xix). He therefore revised Arnold's "to see the object as in itself it really is" to his relativist formula "to know one's own impression as it really is" (TR, xix). This philosophically empirical method, Pater argued, freed the aesthetic critic from idealist conceptions of beauty: "And he who experiences these impressions strongly, and drives directly at the discrimination and analysis of them, has no need to trouble himself with the abstract question what beauty is in itself, or what its exact relation to truth or experience—metaphysical questions, as unprofitable as metaphysical questions elsewhere" (TR, xx).

Pater's acceptance of impressionism produced a central inconsistency in his criticism. The objective and empirical assumptions of scientific investigation which led him to conclude that all human perceptions were subjective also informed his practice of critical historicism. Philip Appleman has pointed out this logical incompatibility: "The horns of the dilemma are his impressionism and his historicism, each of which, to be consistent, requires the exclusion of the other."[90] Appleman argues that Pater's dilemma is intellectually insoluble for the post-Darwinian critic: "[T]his indecision was a natural result of his Darwinian conditioning. . . ."[91] However true this may be for the modern critic, there was a solution—Hegelian dialectic—available to Pater, but as Anthony Ward observes,[92] Pater did not accept Hegel fully until very late in life ("Plato and Platonism," 1893).

Wilde, however, had no doubts about a Hegelian dialectical resolution. Some of his notebook entries on science and poetry explore the dialectical relationship of the subjective, creative faculty and the objective, descriptive faculty, a relationship he later developed in *Intentions*. For instance, he notes that "astrology and alchemy mark the time before poetry and science had been differentiated" (CN, 22). Their original synthetic unity suggests how, as Wilde paraphrases from Froude, "Poets are the original historians as well as the original men of science" (CN, 101). So, Wilde applied dialectical thought to the development and relationship of poetry and science just as he did with history and philosophy. Scientific thought is first conceived imaginatively and then later differentiated from the poetic faculty. In an important entry he notes that the early Greeks "had mystic anticipations of nearly all great modern scientific truths" (CN, 43), and he cites examples from Ruskin's *Queen of the Air* to support this claim. Modern science errs when it excludes imagination, when it is too inductive, too concerned with collecting, observing, and analyzing facts. The solution to this problem lay in resurrecting and developing the poetic faculty in science. A similar analysis appears in the "Rise": Progress requires a return, in a more developed form, to the intimate connection of poetry and science. Like the Greeks, modern scientists could use imagination and emotion as well as observation, experiment, and analysis. Wilde cites conceptions such as the fourth dimension of space, infinity, and eternity as examples of imaginative science (CN, 42).

In several entries Wilde noted the fruitful connection of poetry and science when the "poetic imagination furnishes the principles and so is connected with the deductive method." From Buckle he cited the scientific achievements of Newton, Harvey, Black, and Leslie (CN, 43, 45; see also CW, 1122n.1). He named Shelley as a modern poet fascinated by science who "if he had lived might have given to modern thought this elevating character" (CN, 45; CN, 22; and CB, 183).[93]

Wilde also cited "use of the scientific knowledge in poetry" (CN, 22), but scanted the topic. He gave examples of poets who used science but he found its most significant contributions to be definite conceptions and general ideas which inform and inspire poets—ideas like vastness, law, ensemble, heredity, evolution, force, and progress (CN, 22–25). He also believed that instruments like the microscope and telescope gave poets and scientists new ways to reconceive the world (CN, 31). Later in the notebook Wilde reminded himself to use such information: "For a metaphor • There are stars which send a blue and red light" (CN, 113).

Wilde obviously believed modern civilization overused and overvalued objectivity and devalued emotional, subjective, and imaginative perceptions. However, he did not believe that these faculties were ultimately separable: Scientific perception was informed by imagination, just as imaginative vision could be generated by scientific insight. For Wilde "seeing the object as in itself it is not" included the scientific effort to see it "as in itself it really is." Like his historical method,

Wilde's view of science made the observation, collection, and analysis of knowledge dialectically inductive and deductive: "[A]nd primarily rem. that man must use all his faculties in the search for truth" (CN, 43). In aesthetics Wilde's dialectical method enabled him to transcend Paterian impressionism. Furthermore, Wilde's acceptance of Hegelian metaphysics placed him, almost literally, worlds apart from Pater's intellectual and social ideas. We believe that this intellectual context leads inevitably to a crucial reinterpretation of Wilde's critical position and provides what Richard Ellmann misses in Wilde's critical theory, a metaphysical basis which unifies contradictory ideas and provides an escape from impressionism.[94] Pater influenced Wilde's taste in art, but even in his last critical essays Wilde never gave up the Hegelian humanism he learned at Oxford.

Wilde's idea that "the primary aim of the critic is to see the object as in itself it really is not" (CW, 1030) should not be understood as a witty rephrasing of Pater's impressionism. Rather, it implies the Hegelian recognition that one approaches truth through the apprehension of contraries, that a critic's subjectivity and objectivity are real and can only function in a dialectical relationship. Wilde revised Arnold's original formula—"to see the object as in itself it really is"—by subsuming and assuming Arnoldian objectivity while insisting upon the primacy of the critic's imaginative subjectivity. Wilde fully realized these dialectical functions of criticism in "The Critic as Artist." At the end of the first section, shortly after he describes the subjective-imaginative responsibility of the critic, Gilbert announces that after supper "we will pass on to the question of the critic considered in the light of the interpreter." Ernest replies that if this critic does interpret, he "may occasionally be allowed to see the object as in itself it really is" (CW, 1032). In the second section Gilbert admits this function:

> Yes; the critic will be an interpreter, if he chooses. He can pass from his synthetic impression of the work of art as a whole, to an analysis or exposition of the work itself, and in this lower sphere, as I hold it to be, there are many delightful things to be said and done. . . . Ordinary people are "terribly at ease in Zion." They propose to walk arm in arm with the poets, and have a glib ignorant way of saying, "Why should we read what is written about Shakespeare and Milton? We can read the plays and the poems. That is enough." But an appreciation of Milton is . . . the reward of consummate scholarship. . . . The critic will certainly be an interpreter, but he will not treat Art as a riddling Sphinx, whose shallow secret may be guessed and revealed. . . . Rather, he will look upon Art as a goddess whose mystery it is his province to intensify, and whose majesty his privilege to make more marvellous in the eyes of men. (CW, 1032–33)

Although in his later criticism Wilde favored a creative and subjective mode, he did not believe this mode was epistemologically inevitable. He did not, like Pater, think himself into the prison of personality. His method of criticism, as we see it in his American lectures and in *Intentions,* shows him developing the interactive modes of scholarly and impressionistic criticism. For Wilde the functions of criticism were to see the object as in itself it is *and* is not. This position constitutes a development, not a rejection, of the assumptions which inform the "Rise."

Wilde's Lectures in America

In a series of lectures on the Pre-Raphaelite Brotherhood and the decorative arts movement delivered in North America in 1882, Wilde declared his differences with Walter Pater and his allegiance to John Ruskin. In Wilde's original lectures, based on his intellectual synthesis, he discussed history, crafts, fashionable dress, home decoration, poetry, politics, the novel, and criticism.[95] As critics have noted, Wilde relied heavily on ideas drawn from the writings of Ruskin and Pater but set them within a theoretical context clearly his own.[96] Wilde, appointed in a *Punch* caricature of 1880 to the position of "Professor of Aesthetics and Art Critic," lectured to provide publicity for *Patience,* Gilbert and Sullivan's satire on the aesthetic movement. But Professor Wilde, we think, used his celebrity to offer a serious, academic antithesis to this popular, Philistine interpretation of the movement and to the aesthetic hedonism of Pater's "Conclusion" to *The Renaissance.*[97] Wilde would have recognized Pater's inconsistent adoption of Hegelian philosophy: Pater accepted the broad outlines—and many details—of Hegel's history of art, but on epistemology, logic, and metaphysics he deserted the idealist camp. Wilde differed from Pater on these issues and their implications for art and ethics. In the Commonplace Book and both versions of "The English Renaissance of Art," Wilde showed his differences through clever allusions to, and changes in, Pater's ideas and language in "Winckelmann" and the "Conclusion."[98]

These lectures, Wilde's first important public pronouncements on art and culture, are an extension of the idealist-racialist synthesis developed in the notebooks and the "Rise," an aesthetic philosophy (a kind of Hegelian Ruskinism) which Wilde later revised in *Intentions* and "The Soul of Man Under Socialism." In the lectures Wilde analyses the history and nature of the Arts and Crafts movement from a Hegelian perspective. He follows Hegel, defining "reality" as a concrete universal and believing that the history of art (like all history) is characterized by the progressive realization of consciousness (or spirit) through time. However, Wilde revised Hegel's aesthetic theory by suggesting a more complex view of historical development and by giving greater significance to the material nature of art and culture.

Wilde's historical analysis suggests that the Arts and Crafts movement grew out of concrete historical circumstances and is the newest, most self-conscious manifestation of spirit in art. His discussions of the artist, artworks, and the social function of art also synthesize the concrete and the ideal. The artist is "the child of his own age" and "the spectator of all time and of all existence" (ER, 458). Artworks are unions of realism and the spirit: "[T]hat work is most instinct with spiritual life which conforms most clearly to the perfect facts of physical life" (ER, 449; cf. W, 43). He asserts, furthermore, that art has social functions, but that these are accomplished only if the artist, the artwork, and the audience do not have directly pragmatic concerns. In short, only when we recognize that the artist and art exist in a separate world can their practical value be realized: "Art never harms itself by keeping aloof from the social problems of the day: rather, by so doing, it more completely realises for us that which we desire" (ER, 458).

The metamorphoses of these lectures, as documented by Kevin O'Brien, reveal much about the place and importance of theory for Wilde. Of the three, the first version of "The English Renaissance of Art" is most important for an understanding of Wilde's theory.[99] The critical position he adopts, the social and aesthetic problems he defines, and the terms and explanations he uses show that Wilde shares Ruskin's and Hegel's belief in spiritual seeing and Ruskin's concern for the efficacy of art in promoting individual and social perfection. As in the "Rise," he employs a synthetic and critical Hegelian explanation of history and art. He uses Hegel to resolve the contradictions even Ruskin recognized in his own theories[100] and to synthesize conflicting historical and philosophical movements, producing an evolutionary history of the English Arts and Crafts movement, which he defines as a cultural phenomenon antithetical to dominant English culture. But he disagrees with Pater, and so with Hegel, on two related and crucial matters. First, he breaks with orthodox Hegelian aesthetic theory by emphasizing continuities between, and common characteristics in, Greek and romantic art. Second, Wilde revises Paterian art history by seeing the English renaissance as a rebirth of Hellenic self-consciousness in modern form, that is, Hellenism incorporating certain characteristics of modern romantic art. As in "Hélas," romantic indulgence is not ultimately at odds with his spiritual inheritance.

In "Winckelmann" Pater restates the Hegelian historical analysis that spirit moves from Greek sensuous form to a higher, less sensuous Christian form (the soul) which expresses its own dissatisfaction with the body. According to Hegel, the spirit's realization, even in art, is demonstrated by a progressive detachment from the physical world until it achieves full self-consciousness in philosophy as pure, disembodied mind thinking of itself thinking. Pater restates this Hegelian theory: "It is sometimes said that art is a means of escape from 'the tyranny of the senses.' It may be so for the spectator . . ." (W, 47).[101]

Wilde disagreed with this Christian, romantic, and philosophic notion of spirit as disembodied essence (soul). For him human perfection was expressed in the Hellenic harmony of body and spirit, and progress, as he interpreted it, meant a return to (or recreation of) the Hellenic spirit. Wilde believed that because art required the sensuous embodiment of the spirit, it and not transcendental philosophy best expressed this human ideal: "Yet is this spirit of transcendentalism alien to the spirit of art. For the artist . . . there is no escape from the bondage of the earth: there is not even the desire of escape" (ER, 448–49).[102] In his lecture Wilde shows his differences with Pater and Hegel when he remarks, echoing Pater's prose while rejecting his theory, "for while art has been defined as an escape from the tyranny of the senses, it is an escape rather from the tyranny of the soul" (ER, 475–76; cf. ER, 463).

Wilde appears to contradict himself here. On the one hand he talks of the spiritual essence of art, while on the other he claims art is an escape from the tyranny of the soul. His answer would be that true spirit or soul can only be expressed or felt through the sensuous. This position offers a synthesis of the two partial, and therefore unhealthy, manifestations of spirit in modern life. Like Carlyle, he saw traditional Christianity and skeptical materialism as antitheses. Christianity interpreted spirit as totally nonmaterial, while skepticism, as he said, was to the soul as a shadow is to a body, a negation which proves the existence of what it denies (ER, 475). The healthy spirit recognizes the inextricable relation of body and soul.

Wilde therefore offered art as a cure for the anxiety of materialists like Pater, who doubted the metaphysical reality of the human soul: "[T]here are times when wisdom becomes a burden and knowledge is one with sorrow: for as every body has its shadow so every soul has its scepticism. In such dread moment of discord and despair where should we, of this torn and troubled age, turn our steps if not to that secure house of beauty" (ER, 475). This *consolation des arts* is a reassociation of sensibility which comes from the ability of the senses, operating through *theoria,* to cleanse and restore the soul.[103] He defined spiritual and physical health Hellenically as a golden mean (reinterpreted in Hegelian terms), while art, as the synthetic mode (rather than philosophy or religion), became the source of Wilde's New Hellenism.

Wilde's historical interpretation of the nineteenth-century English renaissance is consistent with this synthetic ideal. He described the movement as a product of "Greek modes of thought" and "mediaeval feeling," to which it added "the intricacy and complexity and experience of modern life" (ER, 444). He characterized the synthesis in an image of marriage borrowed from Pater: "It is really from the union of Hellenism, in its breadth, its sanity of purpose, its calm possession of beauty, with the adventive, the intensified individualism, the passionate colour of the romantic spirit, that springs the art of the nineteenth century in England, as from the marriage of Faust and Helen of Troy sprang the beautiful boy Euphorion" (ER, 444).[104] Wilde adopted the Hegelian division of Western art history into two spirits or ages, the classic (which he also calls Hellenic) and the romantic (which he also calls medieval). His equation of medievalism with romanticism is strange now, but this duality worked remarkably well for dialectical analysis. Ancient and modern, mind and feeling, type and exception, essence and accident, calmness and passion, clarity and mystery, permanent and momentary—all of these oppositions Wilde saw as historical expressions of the human spirit which formed "the essential elements of our conscious intellectual tradition, of our permanent standard of taste" (ER, 445). In a characteristically Hegelian way Wilde posits a single spirit in both art and politics that generates the dualities of history: "[O]ne high law, the law of form or harmony" expresses the human desire for a nobler form of life and for freedom (ER, 445).

In modern history this new Hellenism was reborn as "that critical spirit of Germany and England which accustomed men to bring all things to the test of reason or utility or both" (ER, 446), while its attendant "desire for perfection" (ER, 449) paved the way for the French Revolution, the "primary factor" (ER, 446) or source of the movement. So, for Wilde spirit animates history and produces evolutionary change, and history modifies spirit to produce a new impetus for artistic creation. He refines his analysis by placing the movement in the context of the history of art: He sees Pre-Raphaelite art and literature as opposing previous and more abstract movements such as English eighteenth-century poetry, French neoclassical drama, German sentimental literature, and English and American transcendentalism (ER, 448). The artists' choice to imitate art before Raphael, and their adoption of a Hellenic realism, sense of calm, and ideal of formal perfection arise from a synthesis of the dominant and apparently opposing forces of the nineteenth century—the development of science, which Wilde associates with Shelley, and the love of artistic beauty for its own sake, which he associates with Keats (ER, 461). Wilde's theory of art and the artist, then, reconciles the forces of past and present: "The artist is indeed the child of his own age, but the present will not be to him a whit more real than the past" (ER, 458).

In his notebooks and in this essay Wilde identified science with the "democratic and pantheistic tendency" and, as he had done in the College Notebook, used Shelley as the representative of it (ER, 461, 447; CN, 22–25). For Wilde the Pre-Raphaelite artist was, like a scientist, a "true realist" because of his "love of definite conception," "clearness of vision," and "artistic sense of limit" (ER, 448).[105] Pre-Raphaelite art had an antitranscendental quality, a Hellenic absorption in the sensual world similar to scientific inquiry. However, "realism" for Wilde is only a single aspect of true art, a recognition of spiritual life. Art provides a perfect realization of the concrete universal when it is created by men like Keats, who embody within it spiritual life, personality, imagination, and a vision of the real world.

Wilde identifies the qualities of the second important force in the synthesis—the love of beauty—with Keats, "the pure and serene artist, the forerunner of the pre-Raphaelite school" (ER, 450). The "calmness and clearness of his vision, his perfect self-control, his unerring sense of beauty and his recognition of a separate realm for the imagination" (ER, 450) are all characteristic of "the tendency to value life for the sake of art" (ER, 461). In naming Keats as a founder of the Pre-Raphaelite movement, and therefore a transmitter of the romantic/Hellenic spirit, Wilde again differs with Pater. In "Winckelmann" Pater criticized Keats's "sensuous nature" as a limiting characteristic, one he compared unfavorably to a "complete composite nature" like Shakespeare's, Raphael's, or Goethe's. For Wilde the work of Keats and the Pre-Raphaelite Brotherhood represented a synthesis of romantic and Hellenic qualities. Their Hellenic, self-conscious concern with the selection, control, and arrangement of artistic media—be they stone, paint, or words—tempered romantic excesses such as the symbolist tendency of transcendentalism. The works of the Pre-Raphaelite Brotherhood thus combined romantic energy, passion, and expression of individuality with a Greek concern for formal beauty.

Wilde further complicates his analysis with a discussion of technology and artistic materials. Historically art has improved with every new discovery of a "material of workmanship," which has "in itself incommunicable and eternal qualities of its own" (ER, 455). Art developed from constant dialectical tension between the spirit inherent in the materials of workmanship and the "spirit and the method of the poet's working—what people call his inspiration" (ER, 455). The artistic spirit manifested itself in the artist, the materials of art, and in the artist's choice of subject matter.

Wilde argued that William Morris' art achieved a synthesis which represented the perfection of Pre-Raphaelite art because it embodied the quest for the ideal manifested as an exquisite sense of design and spiritual vision, in short, *theoria*.

Without a hint of didacticism or moralism, Morris has "given to our individualised romantic movement the social idea and the social factor also" (ER, 453). Morris' decorative art contained eternal truths that directly affected the perceivers, thus resolving the apparent contradiction that art should remain aloof from social problems but should also have a social function. Wilde believed that spirit, in the form of the artist's aesthetic sense, is present in all true art but is felt only indirectly, through its medium, and is never represented by an artist's choice of subject or didactic comment. However, Wilde's position has been misunderstood by modern readers as a commitment to art for art's sake. His famous remark, "[O]ne should never talk of a moral or an immoral poem—poems are either well written or badly written, that is all" (ER, 470; cf. ER, 2, 113–14; CW, 17) appears first in this lecture and has been taken as an affirmation of that movement. However, the actual significance of the remark is to discourage explicitly didactic art. Art's social function can only be indirect and utopian, since it "more completely realises for us that which we desire" in an age when "to most of us the real life is the life we do not lead" (ER, 458). That is, realized formal perfection provides a model for the perfection which we desire in life. In short, true art will have an important social influence, but it will be provided by the work itself or by the critic.

Wilde stipulates that critics should neither lecture to artists about what is right or wrong in their work nor explain to the public the essentially mysterious aesthetic experience created by art. However, the critic should educate the public about the history of, the appropriate conditions for, and the responses to art: "it is for the critic to create for art the social aim, too, by teaching the people the spirit in which they are to approach all artistic work, the love they are to give it, the lessons they are to draw from it" (ER, 465–66). Like Ruskin, Wilde criticizes "the commercial spirit" (ER, 466) which has destroyed sculpture by (in Gautier's phase) killing the visible world; it has similarly killed drama by creating an ignoble public, lacking the "great national united energy" and the "lofty moral and spiritual ardour" of Elizabethan England and Periclean Athens (ER, 467–68).[106]

Wilde also believed that when the critic concerned himself with beauty he was improving society. He theorized that by encouraging the public to study and appreciate art and to create beauty in their homes, their cities, and nature, the critic would help the Anglo-Saxon race[107] to develop a sorely needed aesthetic sense through the inheritance of acquired characteristics: "I know, indeed, that the divine natural prescience of beauty which is the inalienable inheritance of Greek and Italian is not our inheritance" (ER, 473). Without such a critical, aesthetic sense, nations and individuals waste their strength, follow false ideals, and, worst of all, mistake feeling for form (ER, 470), whereas the "devotion to beauty and to the creation of beautiful things is the test of all great civilised nations" (ER, 471). Furthermore, Wilde reminded his American audience, art's influence transcends national interests: "[A]rt is what makes the life of each citizen a sacrament and not a speculation, art is what makes the life of the whole race immortal" (ER, 471). Wilde's use of sacrament as metaphor precisely defines his understanding of art's inheritable social function: The experience of art enhances mind and spirit, and the cumulative effects are passed on to subsequent generations. In this way each citizen bequeathes immortality to the race. "For the good we get from art," he says, "is not what we learn from it; it is what we become through it.[108] Its real influence will be in giving the mind that enthusiasm which is the secret of Hellenism, accustoming it to demand from art all that art can do in rearranging the facts of common life for us . . . [and] in accustoming it to love the things of the imagination for their own sake, and to desire beauty and grace in all things" (ER, 477). Once the sense of beauty has developed, Wilde believes, people will be unhappy unless beauty is manifest in all of life: "For he who does not love art in all things does not love it at all, and he who does not need art in all things does not need it at all" (ER, 477). In short, Wilde believes that the principle of beauty will develop and operate in art much as the principle of uniformity has in science (CN, 105).

Following Ruskin, Wilde claimed that the creation and purchase of "what gave pleasure to the maker of it and gives pleasure to its user" (ER, 476) would enable both craftsmen and consumers to live " 'in a simple atmosphere of all fair things' " (ER, 474). As Plato taught Wilde, art could then exercise its powerful humanizing function by providing a beautiful environment which would " 'insensibly and gradually draw the child's soul into harmony with all knowledge and all wisdom' " (ER, 474).[109]

Wilde also differed with Pater about art's social power and the critic's function, as is evidenced in a series of Commonplace Book entries on the conclusion to "Winckelmann." Wilde could not accept Pater's materialist assumptions about reality or his beliefs about modern culture. In "Winckelmann" Pater incorporated an idealist history of art within a materialist, individualistic, utilitarian interpretation of culture and society. His philosophical and political values appear in the conclusion where he discusses art's role in producing individual "self-culture" (W, 49), heightened self-consciousness, or Aristotelian *energeia*. As we have argued, he applies the utilitarian model of economic man to this process of self-culture: humans are consumers who attain success in life by maximizing their aesthetic pleasure. Pater's materialism is evident in his comment that "the chief factor in the thoughts of the modern mind concerning itself is the intricacy, the universality of natural law even in the moral order" (W, 50), and in his belief that moderns should renounce "a taste for metaphysics" and "the fancied gift of absolute or transcendental knowledge." Instead, he argued, modern art could best serve culture by providing "an equivalent for the sense of freedom" that deterministic science has destroyed (W, 50).

In the Commonplace Book Wilde writes that Pater's practice as an art critic contradicts his announced renunciation of metaphysics and transcendental knowledge: "Yet surely he who sees in colour no mere delightful quality of natural things but a spirit dwelling in things is in a way a metaphysician" (CB, 169).[110] Wilde also recognized (in an entry on the facing page) that people who dwell only in the realm of pure metaphysics are equally in error (CB, 168). He opted, in short, for spirit as it is embodied in the sensual, for art rather than philosophy, and this position prefigured his revision of Hegelian art history in "The English Renaissance."

In another entry, Wilde disagreed with Pater's notion of "what modern art has to do in the service of culture'' (W, 50). Pater adopted Hegel's idea that music and poetry are the representative arts of the modern world because they provide ''the sense of freedom'' which ''the spirit'' needs ''in the face of modern life'' (W, 50). Philosophical materialism provided Pater's escape from a Puritan, repressive religion, but it also created a materialist malaise (the absence of freedom) that he believed art could alleviate. Wilde accepted Pater's (and Hegel's) choice of representative arts, which ''may be regarded as the natural expression of the spirit of man projecting itself in the forms of beauty most characteristic of its nature,'' but he gave both poetry and music a different function, to provide ''the sense of the infinite, and the scientific spirit of criticism to modern life'' (CB, 163). Wilde, using a Hegelian idealist notion of spirit, suffered no materialist malaise. The modern spirit is not limited by a loss of freedom; rather, it can synthesize and express in art infinite yearnings as well as critical reflections.

Like Pater, Wilde understood the value of scientific determinism; however, unlike Pater, Wilde recognized its theoretical limitations: It was a useful but partial point of view. Wilde continued his dialogue with Pater on this subject in another Commonplace Book entry, where he records Pater's metaphorical description of necessity as a ''magic web'' woven through and through us like a network of nerves, which extends ''the universality of natural law even in the moral order'' (W, 50). That is, Pater believed the modern mind could no longer accept the conception of a free, moral will. Art, however, could replace freedom with feelings of blitheness and repose to compensate for this loss. Wilde responded to Pater's denial of free will in two sentences which embody his own idealist position on necessity: ''When we combat it, [it is] with the weapons of its own forging, and the assertion of liberty is only the claim that from certain forms of conscious volition certain results will inevitably follow• We are indeed compassed by the high . . . mountains of necessity, but for him who knows his limitations this dark horizon becomes the sunlit circle of duty'' (CB, 171). Wilde's answer to the materialist malaise is a synthesis of science and idealism: Liberty exists as a conscious volition which materialists cannot exclude from the chain of necessity, and, social responsibility (duty) can therefore not be denied as soon as liberty is acknowledged. It is, after all, important to be earnest (as long as one does it in the right spirit).

At the end of ''The English Renaissance,'' Wilde openly differed with Paterian materialism by including yet transcending (as he had in the notebook entries on ''Winckelmann'') the egoistic hedonism recommended in the ''Conclusion'' to *The Renaissance*. Wilde wants ''to dwell on the effect that decorative art has on human life—on its social not its purely artistic effect'' (ER, 478). He describes the value of the renaissance in art for the ''two kinds of men in the world . . . men to whom the end of life is action, and men to whom the end of life is thought'' (ER, 478). Then Wilde paraphrases Pater's ''Conclusion'' to describe the function of art for anyone ''who must burn always with one of the passions of this fiery-coloured world'': ''For art comes to one professing primarily to give nothing but the highest quality to one's moments, and for those moments' sake'' (ER, 478). However, Pater's conclusion is not conclusive for Wilde: Art can do more than Pater claims, and the critic with the theoretic faculty should be concerned with more than self-culture.

Against Pater's individualistic hedonism, Wilde proposes a social function of art for laborers: "[T]o them," he says, "should this movement be specially dear: for, if our days are barren without industry, industry without art is barbarism" (ER, 478). Concluding his lecture, Wilde discusses art as a social institution which could humanize the workers who produce it and the people who own it. Reasserting the causal relations of environment and mind, he insists (quoting Morris) that the true test of a workman is neither attitude nor productivity but "his power of design," which " 'is the studied result of accumulative observation and delightful habit.' All the teaching in the world is of no avail if you do not surround your workman with happy influences and with beautiful things" (ER, 479–80).

Wilde's lecture ends gnomically with the assertion that "the secret of life is in art" (ER, 481). To his audience the phrase might have meant the revaluation upward of art's place in culture. To Wilde it might have suggested his own life in art, the adoption of the role, humorously announced in *Punch,* of critic and artist. (In 1882 he had recently published his volume of poems and hoped to become celebrated as a lecturer who delivered serious ideas in distinctive and aphoristic form). In another sense, though, the phrase suggests Wilde's idealist theory of art and interpretation: Art's spiritual secrets reflect the artist's vision mediated by the potentialities of form. Critics, readers, and viewers can understand these secrets if they interpret with *theoria.* There is no overt didacticism in such creation and response, but individual lives are improved by their contact with the beautiful, and so, through inheritance of culture, are generations of descendants. Wilde later expanded the social implications of this theory in "The Soul of Man Under Socialism." Here, in "The English Renaissance of Art," he is more concerned to state his Hellenic-Hegelian-Ruskinian aesthetic theory and to offer his audience a dialectical revision of Pater's practice, a praxis that, by incorporating art into life, could make life into an art.

Intentions and the Dialectical Method: "The Truth of Masks"

Theorists and historians of criticism have rightly singled out *Intentions* (1891), Wilde's collected major criticism, as the most comprehensive expression of his thought. Most would agree with William K. Wimsatt, Jr., and Cleanth Brooks when they call *Intentions* "perhaps the most imposing theoretical monument" of the art-for-art's-sake movement.[111] René Wellek asserts that in *Intentions* Wilde "merely restates with greater brilliance and wit the main ideas he had absorbed from Pater."[112] Wellek points to Wilde's idealist aestheticism, his belief in the autonomy of art, as his most important critical contribution: "In his finest moments Wilde grasps the dialectics of art: the union of content and form, technique and personality, the subjective and the objective, the conscious and the unconscious, and sees that only in art criticism can we 'realize Hegel's system of contraries' " (HMC IV, 413). Wellek asserts that Wilde intermittently and inconsistently argued three critical positions: "Pan-aestheticism, the autonomy of art, and a decorative formalism" (HMC IV, 409). Wellek is one of the few critics to notice Wilde's use of Hegel, but he only partially detects the influence of idealism in Wilde's aesthetics. Like other critics, he misses the unity of Wilde's critical and cultural theories.[113] For these reasons he does not recognize the intellectual coherence of *Intentions*.

The consistency of Wilde's critical theory and its development from the "Rise" and the American lectures appears in both the argument and the revisions of the first essay of *Intentions,* "The Truth of Masks." Wilde's assumptions about the relationship of past and present, and his terms for analyzing drama (historical accuracy and artistic form) express the critical and creative (i.e., the inductive and deductive) categories of mind which inform the "Rise." In "The Truth of Masks" he provides authority for his solution to a contemporary aesthetic problem by describing a less developed and self-conscious historical precedent: Shakespeare's creative use of Renaissance historicism gives him a model for modern dramatic production, just as Greek criticism provided the prototype for modern historical methodology. The essay appeared in two versions: first in the May 1885 issue of *The Nineteenth Century* as "Shakespeare and Stage Costume" and then in *Intentions,* with revisions of phraseology, significant emendations, and additions. It is the earliest essay in the collection and the simplest of Wilde's attempts to write dialectical criticism. A few crucial changes Wilde made in the later version show his development of a dialectical form unifying creative and critical functions.

The argument of both versions remains essentially the same until the last paragraph. Wilde begins by advocating a modern idea of theatrical illusion appropriate to an age fascinated by history and science.[114] The title of the essay accurately describes its subject—costume—as a revelation of truth. Wilde assumes the role of an objective historian of drama to interpret the historical and artistic truths inherent in Shakespeare's use of costume: "Costume is a growth, an evolution, and a most important, perhaps the most important, sign of the manners, customs and mode of life of each century. . . . A historian who disregarded it would give us a most inaccurate picture of the time, and a dramatist who did not avail himself of it would miss a most vital element in producing an illusionist effect" (CW, 1074).

Wilde shows that Shakespeare's drama reflects the characteristic Renaissance interest in studying and interpreting the past: "[A]nybody," he says, "who cares to examine the age of Shakespeare will see that archaeology was one of its special characteristics" (CW, 1066). He also asserts that Shakespeare was an archeologist, citing the dramatist's attentive fidelity to authentic historical documents, folk history, and individual and racial character. Wilde concludes that "not merely did he select fact instead of fancy as the basis of much of his imaginative work, but he always gives to each play the general character, the social atmosphere in a word, of the age in question. . . . He is even true to the characteristics of race. Hamlet has all the imagination and irresolution of the Northern nations, and the Princess Katharine is as entirely French as the heroine of *Divorçons*. Harry the Fifth is a pure Englishman, and Othello a true Moor" (CW, 1070). Wilde was not advocating archeology for archeology's sake; rather, it is "a method of artistic illusion, and . . . costume is a means of displaying character without description, and of producing dramatic situations and dramatic effects" (CW, 1078). He supports his argument with copious quotations from Shakespeare's plays which suggest that Shakespeare "saw that costume could be made at once impressive of a certain effect on the audience and expressive of certain types of character, and is one of the essential factors of the means which a true illusionist has at his disposal" (CW, 1065). The value of archeology "depends entirely on how it is used, and only an artist can use it. We look to the archaeologist for the materials, to the artist for the method" (CW, 1069). Because Shakespeare is the paradigmatic artist who transforms archeological facts to aesthetic forms, Wilde concludes that by mounting a play with historically accurate scenery and costumes "we are carrying out Shakespeare's own wishes and method" (CW, 1072).

Shakespeare's historicist intentions, Wilde implies, embodied the characteristic development of dramatic art. It is a principle, Wilde claims, that "Art has no other aim but her own perfection, and proceeds simply by her own laws" (CW, 1074). This is not an affirmation of art for art's sake but Wilde's Hegelian description of the spirit, or *Geist,* of art working itself out in history. Wilde insists that "the aesthetic value of Shakespeare's plays" depends neither on beauty nor facts "but on their Truth, and Truth is independent of facts always, inventing or selecting them at pleasure" (CW, 1071).[115] Charles Taylor, a modern student of Hegel, affirms Wilde's understanding of the Hegelian view of art: "Now when the absolute, *Geist,* or the truth that the world emanates from Spirit, is presented out there in sensuous reality, we have the domain of art. That is what art is, in Hegel's view. No matter what the subject of a work of art is in the ordinary sense, the quality of beauty, that which draws us and delights us, comes from the manifestation of the Idea which it embodies in sensuous form. Beauty is the 'sensible manifestation of the Idea.' . . ."[116] Wilde's clearest statement that the revelation of the Idea is the foremost function of art appears in his definition of "the true dramatist": "[I]t must be remembered that neither in costume nor in dialogue is beauty the dramatist's primary aim at all. . . . The true dramatist, in fact, shows us life under the conditions of art, not art in the form of life" (CW, 1075). Through historical scholarship Wilde argues for Shakespeare as a true dramatist, one who used archeological detail as a means to "touch the dry dust of antiquity into the very breath and beauty of life . . ." (CW, 1067). That is, Shakespeare was not only concerned with the creation of Truth in art but with the appropriate means—the grammar—of communicating it to his audience.

In the second half of the essay Wilde argues that historically accurate settings and costumes are equally justified because they are an appropriate grammar for the modern audience, which, like its Renaissance counterpart, has a historical consciousness: "Racine produced his Roman plays in Louis Quatorze dress . . . but we require different conditions for the enjoyment of his art. Perfect accuracy of detail, for the sake of perfect illusion, is necessary for us" (CW, 1073). In short, he insists that any dramatic work "should be presented to us in the form most suitable to the modern spirit" (CW, 1073). It follows, Wilde argues, that the illusion created by archeological detail has its greatest effect on the modern audience when the historical accuracy is sustained by an artistically unified production. "In fact, in art there is no specialism, and a really artistic production should bear the impress of one master, and one master only, who not merely should design and arrange everything, but should have complete control over the way in which each dress is to be worn" (CW, 1077). While this plea for artistic unity may seem commonplace today, it implies Wilde's rejection of the Romantic and mid-Victorian glorification of individual actors who used or abused plays for their own ends. Instead Wilde wished to substitute, carefully orchestrated ensemble acting.

This section on production for a modern audience concludes the argument of the original essay (1885). In it, Wilde practices the theory he developed in the notebooks, from the emphasis on accurate historical understanding, to the need for "one master" interpreter, presumably endowed with *theoria*. He shows his commitment to both archeological detail and formal artistic unity (facts and the spirit that selects and shapes them) because both are necessary to create "Truth" in art. Furthermore, the structure of the historical argument (a moment of the past recognized as a model relevant to present historical and artistic practice) repeats he Hegelian pattern of the "Rise."

Before the publication of *Intentions* Wilde made a few small but important changes and added a coda to the final paragraph. All these adjustments emphasize and clarify the Hegelian strain in the essay, even as they transform it, making it an embodiment of dialectical consciousness. They therefore demonstrate Wilde's intentional transformation of idealist historical criticism into what he called "art-criticism," a form of discourse which fuses subjective and objective modes, creative-critical functions. The small changes occur for two purposes. First, they eliminate many of the topical aspects of the controversy and make the essay a more general argument. Second, they refine the terminology of the essay to make it consistent with Wilde's idealist stance. For instance, Wilde added the subtitle, "A Note on Illusion," and in the text he altered all descriptions of good drama and dramatists from "realistic" to "illusionistic." In the last paragraph of the essay Wilde changed "a method of realism" to "a method of artistic illusion" in the following phrase: "What I have tried to point out is that archaeology is not a pedantic method, but a method of realism. . . ."[117] These changes might suggest that Wilde simply disliked realism and preferred the purely aesthetic value of art. But other changes show that Wilde replaced the terms "realism," "realistic," and "absolute reality" with variations of "illusion" to eliminate a possible confusion between the ordinary use of "real," implying empirical reality, and its Hegelian meaning.

Wilde emphasized the idealist content of the essay by changing "truth" to "Truth" in the statement "Truth is independent of facts always, inventing or selecting them at pleasure" (CW, 1071). Elsewhere he clarified his substantive change: In arguing that art makes archeology beautiful, Wilde first said that only art "can combine in one exquisite presentation absolute reality with the grace and charm of the antique world" ("Shakespeare and Stage Costume," 808). In *Intentions* he changed this to read that only art "can combine in one exquisite presentation, the illusion of actual life with the wonder of the unreal world" (CW,

1068). Clearly Wilde did not drop "absolute reality" because he rejected idealism, but because the phrase might be understood as a reference to the reality of archeological facts. Given the direction of the argument, Wilde simply recognized that the term "real" (and its variants) would create more confusion than clarification.

Finally, the philosophical content of "The Truth of Masks" appears most clearly in the coda Wilde added to the last paragraph of the essay:

> Not that I agree with everything that I have said in this essay. There is much with which I entirely disagree. The essay simply represents an artistic standpoint, and in aesthetic criticism attitude is everything. For in art there is no such thing as a universal truth. A Truth in art is that whose contradictory is also true. And just as it is only in art-criticism, and through it, that we can apprehend the Platonic theory of ideas, so it is only in art-criticism, and through it, that we can realise Hegel's system of contraries. The truths of metaphysics are the truths of masks. (CW, 1078)

While this language is both dense and apparently confusing, we believe it provides Wilde's clearest statement about the nature and purpose of art and criticism. It expresses a mature understanding of the meaning and relation of philosophical contradictions in self, art, and criticism and announces a synthetic form, "art-criticism," a mode of discourse consistent with Wilde's philosophical beliefs.

The first four sentences introduce self-contradiction as a reality pervading truth, art, and the author. Wilde dramatizes this notion by asserting his disagreement with himself, and then explains its necessity by arguing that the essay "simply represents an artistic standpoint." Because truth perceived through art can only be partial—in Hegelian philosophy universal Truth is not available through art but only through logic—the closest approximation to universal Truth will be a contradiction: "A Truth in art is that whose contradictory is also true." Wilde's disclaimer creates an antithetical position and as such constitutes a negation which refashions the essay from a scholarly analysis containing a Hegelian interpretation of art history to a creative and critical mode embodying dialectical form.

The final two sentences, then, describe the potential of this creative-critical mode. First, "only in art-criticism and through it" in both form and content can we apprehend the unchanging realm of Platonic ideas. At the same time, Wilde claims, "art-criticism" allows us to "realise" the antithesis (within the history of philosophical idealism) to Plato's static reality, that is, "Hegel's system of contraries," which defines the universal as Becoming, not Being. If the ideal Truths of Plato and Hegel, that is, static and dynamic realities, are the truths of "art-criticism," then, "the truths of metaphysics are the truths of masks." While this sentence seems deliberately gnomic, we believe it provides a summary statement relating the scholarly content of the essay to the formal and metaphysical play of Wilde's coda, which turns the essay into a creative effort; from an empirical perspective it fictionalizes, and so aestheticizes, his arguments. The scholarly discussion of costume as essential to dramatic illusion is therefore transformed into a creative illusion. Since Wilde partially disclaims the first position and offers a second, both become dialectical illusions—masks—which approximate the Truth.

The essay itself must therefore be understood as a dialectical and critical artwork, an illusion which embodies opposing truths. However, creative illusion, in Wilde's Hegelian theory, is more real than simple objectivity. Wilde's transformation of the essay into a "Note on Illusion" should be understood not as a retreat to aesthetic subjectivity but rather as an effort to create "art-criticism," a form which assumes and embodies the interpenetration of the subjective and the objective in reality.

Decay and Progress in "The Decay of Lying"

Wilde was not totally satisfied with "The Truth of Masks" and declined to include it in the French edition of *Intentions,* called *Opinions de Littérature et d'Art,* replacing it with "The Soul of Man Under Socialism."[118] The coda to "The Truth of Masks" is, after all, an afterthought designed to refashion a scholarly essay into "art-criticism." But this minor piece reveals Wilde's intentions for the form and content of the collection's two major essays, "The Decay of Lying" and "The Critic as Artist," in which he fully realized the critical categories and philosophical approach of "art-criticism." In them Wilde adopted the dialogue because it embodied the dialectic, and perhaps because it is associated with Plato, the philosopher-artist who created noble fictions and dramatic illusions to convey truth, and so was one of the first to write "art-criticism."

In her analysis of "The Decay of Lying" Hilda Schiff argues that Wilde was essentially a subjective formalist who adopted the dialogue form because of his "deeply ingrained dialectical habit of mind." Although he does not recognize Wilde's debt to Hegel, Bruce Bashford extends this line of analysis in important and useful ways: He notices that Wilde's critical theory assumes that "the subjective and the objective are linked in a reciprocal, or better perhaps, reflexive relation" and he explores Wilde's development of a subjectivist critical method which, he contends, "works largely by opposition" and produces intellectual growth through its application.[119] Practicing this form of creative criticism enhances and makes self-conscious the critic's instinctive "beauty sense" and so encourages the development of a subjective and expressive comprehension of the world.

Bashford's interpretation is limited because it focuses mainly on "The Critic as Artist" and on the critic's use of art for self-cultivation and self-expression, and, second, because of its subjective-formalist thesis that the systematic "relation between expression and form . . . is the basic principle of Wilde's criticism" (Bashford, 219). By now it should be clear that Wilde differed with Pater on matters metaphysical and was not an epistemological subjectivist. Schiff's and Bashford's thesis that Wilde only develops Pater's theory is placed in proper perspective by Vivian's remark in "The Decay of Lying": "Art never expresses anything but itself. This is the principle of my new aesthetics; and it is this, more than that vital connection between form and substance, on which Mr. Pater dwells, that makes basic the type of all the arts" (CW, 987). Vivian explains that art, viewed historically, "develops purely on her own lines. She is not symbolic of any age. It is the ages that are her symbols" (CW, 987). Cyril agrees and extends the argument: "The spirit of an age may be best expressed in the abstract ideal arts, for the spirit is abstract and ideal" (CW, 988). These remarks suggest that Wilde's differences with Pater only begin with metaphysics: Wilde's Hegelian and evolutionary assumptions led him to analyze both art and criticism as part of a larger cultural process.

In "The Decay of Lying," one of Wilde's wittiest pieces, he describes the role of critical activity in the creative process, of creativity in the critical process and the social implications and effects of both art and criticism on writers and the general populace. To Henry Luce Wilde wrote, "I should like to know your views on my new theory of art contained in ["The Decay of Lying"].[120] Shortly after its publication Wilde commented in a letter to Violet Fane that "The Decay of Lying" was "meant to bewilder the masses by its fantastic form; *au fond* it is of course serious."[121] We take Wilde seriously as a theoretician with a system, so we shall focus our discussion on the serious center of the dialogue, the union of evolutionary theory and idealism; then we shall turn to its creative and playful aspects, which Wilde, punning on his name, uses to be<u>wilde</u>r and beguile, to lead by misleading. Wilde's reputation as a critic has at times suffered because of this creativity. Some critics have misread his misleading comments and dismissed his comic exaggerations as "various studied absurdities" when, in fact, they are seriously comic, models of criticism as in itself it is not.[122]

In compiling the notebooks Wilde learned that Hegel's and Spencer's theories share with Aristotle's a belief that human, social, and biological organisms are guided in their development by internal plans. Both nineteenth-century theorists explained existence as development, as the realization of a potential inherent in the material universe. For Hegel the realization of intentions (or of the idea) in human history occurred through the negation of present existence by antithetical ideas. Spencer's and Clifford's theories of the inheritance of acquired characteristics described evolution as a process of perfection impelled by the inner need of an organism to adapt to its environment. Victorians interpreted both theories as explanations for social progress, for the development and eventual triumph of intelligence or mind.

In "The Decay of Lying" Wilde claims that art, a manifestation of intelligence, performs just this function in life: "Scientifically speaking, the basis of life—the energy of life, as Aristotle would call it—is simply the desire for expression, and Art is always presenting various forms through which expression can be attained" (CW, 985). By identifying energy as the basis of life, Wilde inverted the materialist assumptions of science and so reconciled it with Hegelian idealism. As in the notebooks and American lectures, Wilde rejected Paterian impressionist theory based on *energeia*. Energy is matter's unselfconscious mind desiring expression. Art, the product of *theoria*, or self-conscious mind, provides the form to express life's energy. Nature by itself, Vivian argues in a comment which gives the collection its title, will not create the variety or perfection which is its potential: "Nature has good intentions, of course, but, as Aristotle once said, she cannot carry them out" (CW, 970).[123] Because life is unselfconscious, it is dominated by an "imitative instinct" (CW, 992). Progress occurs only when life is shaped by creative, nonmimetic art (i.e., art that is not itself dominated by the imitative instinct): In this way life realizes its potential and progresses from the realm of necessity to the realm of self-consciousness and freedom. Human life, in this scheme, is the synthesis of Nature and Mind and has the capacities of both. This idealist evolutionary theory provides substance for Wilde's witty thesis that life should imitate art rather than art imitating life.

Vivian's theory is held together by Wilde's synthesis of three terms: nature, art, and life. In an early comment to Cyril, Vivian offers two closely related definitions of nature:

> If we take Nature to mean natural simple instinct as opposed to self-conscious culture, the work produced under this influence is always old-fashioned, antiquated, and out of date. One touch of Nature may make the whole world kin, but two touches of Nature will destroy any work of Art. If, on the other hand, we regard Nature as the collection of phenomena external to man, people only discover in her what they bring to her. She has no suggestions of her own. Wordsworth went to the lakes, but he was never a lake poet. He found in stones the sermons he had already hidden there. (CW, 977–78)

In each of these definitions Wilde separates nature from the mind and its creations. Nature is thus the biologists' "natural, simple instinct" or the epistemologists' "collection of phenomena" that man perceives. Art, therefore, becomes "self-conscious culture," the product of man's mind, or *theoria,* the process of imaginative perception itself. Wilde completes this relationship in remarks he added to the final version of the dialogue. In response to Cyril's challenge to explain how nature imitates art, Vivian, echoing Ruskin, replies:

> Consider the matter from a scientific or a metaphysical point of view, and you will find that I am right. For what is Nature? Nature is no great mother who has borne us. She is our creation. It is in our brain that she quickens to life. Things are because we see them, and what we see, and how we see it, depends on the Arts that have influenced us. To look at a thing is very different from seeing a thing. One does not see anything until one sees its beauty. Then, and then only, does it come into existence. (CW, 986)

This explanation invokes an idealist monism similar to Clifford's, which combines, as Vivian insists, the scientific and metaphysical points of view. While in the earlier definition he sees reality as matter separated from mind, here, in a sense, it is mind that matters since without it nothing is real. Existence is purely ideal because "one does not see anything until one sees its beauty. Then, and then only, does it come into existence." Artists perceive and create beauty and people can perceive this reality/beauty if art shapes their perceptions. The truth for Wilde is, as always, dialectical.

Just as there are two related definitions of nature, so there are two definitions of art. Art as "self-conscious culture" is the created work, the sensuously realized embodiment of mind, while as a process of perception it is the imaginative act itself. Vivian's definitions of art make this distinction (and relation) clear: "Art itself is really a form of exaggeration; and selection, which is the very spirit of art, is nothing more than an intensified mode of over-emphasis" (CW, 978). Art is exaggeration, and the "spirit of art," the Hegelian *Geist,* criticizes or selects.

Both functions involve overemphasis, but the former creates and the latter excludes. Earlier in the dialogue Vivian defines the same creative and critical processes as ''imagination or fancy, or cultivated blindness'' (CW, 970).

Furthermore, art has a crucial function in the progress of culture. As Vivian says, it is ''our spirited protest, our gallant attempt to teach Nature her proper place'' (CW, 970). Modern European culture, like untutored nature, is dominated by an ''imitative instinct'' (CW, 992). To insist upon art as negation of present reality, Wilde wittily refers to the creative process as ''lying'' and laments its decay because it means the loss of imaginative power and the consequent dominance of the imitative instinct in literature and society. Wilde believed individuality created progress and saw conformity as a threat to it: ''It is a humiliating confession, but we are all of us made out of the same stuff . . . that dreadful universal thing called human nature. Indeed, as anyone who has ever worked with the poor knows only too well, the brotherhood of man is no mere poet's dream, it is a most depressing and humiliating reality'' (CW, 975). Elsewhere Vivian explains that ''we are a degraded race, and have sold our birthright for a mess of facts'' (CW, 977).[124] Later he elaborates on themes introduced in the notebooks, lectures, and the ''Rise'':

> Facts are not merely finding a footing-place in history, but they are usurping the domain of fancy, and have invaded the kingdom of Romance. Their chilling touch is over everything. They are vulgarising mankind. The crude commercialism of America, its materialising spirit, its indifference to the poetical side of things, and its lack of imagination and of high unattainable ideals, are entirely due to that country having adopted for its national hero a man who, according to his own confession, was incapable of telling a lie. . . . (CW, 980)

Vivian also explains how this decay has occurred. The history of the decorative arts in Europe, he says, shows the conflict between idealism and materialism, romance and realism, that is, between what he calls ''Orientalism,'' which is imaginative, abstract, and rejects mimesis, and the ''Teutonic'' or imitative spirit.[125] His history of art, echoing ''The English Renaissance,'' adopts but modifies Hegel's description of the dialectical progress from the symbolic art of Egypt, India, and China to the classical and romantic art of Europe. Where Hegel sees an uninterrupted realization of idea in art, culminating in early nineteenth-century romanticism, Vivian, focusing closely on the present cultural situation of English art, sees a decline into realism which threatens its existence: ''if something cannot be done to check, or at least modify, our monstrous worship of facts, Art will become sterile and beauty will pass away from the land'' (CW, 973).

Wilde objects to mid-Victorian realism because its technique (like untutored nature's)—simple copying—reproduces the type, while true art encourages human improvement through exaggeration. As a result, nineteenth-century mimesis has triumphed over true art, producing the realistic novel, which ''presents us with dull facts under the guise of fiction'' (CW, 972), and melodrama, which Wilde paradoxically sees as the epitome of realism's failure, ''the inevitable result of [the] substitution of an imitative for a creative medium'' (CW, 979). Realists like Zola mistakenly try to turn art to immediate and practical use by describing hideous social conditions. Yet this imitative process provides life with nothing new; what life needs for improvement, given its imitative instinct, is unrealistic, impractical, and beautiful models. To be truly useful the artist's imagination must intervene in the imitative and reproductive process to provide new ideals for life: "A great artist invents a type, and Life tries to copy it, to reproduce it in a popular form, like an enterprising publisher" (CW, 982). English Renaissance drama, for example, "enlisted Life in her service, and using some of life's external forms, she created an entirely new race of beings . . ." (CW, 978). Such romantic art will do more, says Vivian, to "improve the conditions of the race" than short-term, practical ameliorations (CW, 983). To create that art, claims Vivian, "it is our duty . . . to revive this old art of Lying" (CW, 990).

The Greeks, he says, understood the influence of art in life because they placed statues of Apollo or Hermes in the bride's chamber "that she might bear children as lovely as the works of art that she looked at in her rapture or her pain" (CW, 983). Here art literally intervenes in the reproductive process. Realism has the reverse effect: "The boy-burglar is simply the inevitable result of life's imitative instinct. He is Fact, occupied as Fact usually is, with trying to reproduce Fiction, and what we see in him is repeated on an extended scale throughout the whole of life" (CW, 983). Obviously Wilde plays with this paradox, yet he is also completely serious in extending the theory of the inheritance of acquired characteristics to explain the influence of art.

As a *jeu d'esprit* Wilde added to the final version of the dialogue an exaggeration of this theory that exemplifies his critical spirit. Cyril challenges Vivian to prove that not only life but also nature imitates art. After Vivian's brilliant demonstration, he inquires of Cyril, "[H]ave I proved my theory to your satisfaction?" Cyril replies, "You have proved it to my dissatisfaction, which is better" (CW, 987).

Like earlier Victorian sages, Wilde enacts the ideas he recommends. In this essay, as in "The Critic as Artist" and "Pen, Pencil and Poison," his play and creativity appear in his dramatic exposition and in his rhetorical tactics of exaggeration and reversal. His critique of the dominant "Teutonic" culture is offered (in dialogue form) by the irrepressible and irresponsible Celtic negation, Vivian. Rodney Shewan has pointed out that both Vivian and Cyril "systematically contradict and misappropriate the accepted meanings of words at crucial points in the argument" (Shewan, *Art and Egotism,* p. 97) as part of an overall strategy, familiar to readers of Wilde, which attacks intellectual and social pieties and myths by turning them rhetorically against themselves. Wilde plays seriously with the contradictions of the dominant culture. He applies conventional moral terms to the aesthetic sphere (e.g., self-sacrifice, sincerity, duty), while he adopts the hedonistic vocabulary of associationism and utilitarianism (e.g., egoism and pleasure-seeking) as moral ideals. Wilde's renaming of the creative process as "lying" exaggerates its subversive nature (reasserting a Blakean awareness of the poet's allegiance to the devil) and, at the same time, allows him to parody and puncture heavily moralized tales of success and failure. Vivian's story of "many a young man" with the "natural gift for exaggeration" who "falls into careless habits of accuracy'' or into consorting with the ''aged and well-informed'' is but one of the best examples of these rhetorical tactics (CW, 973).

Wilde also uses the dialogue to upset conventional expectations, as, for example, when the conversation focuses on Vivian's article ''The Decay of Lying: A Protest.'' Cyril's remarks provoke Vivian's elaboration, and gradually Cyril is converted, though not without some dissatisfaction—to be dissatisfied with things is, of course, an important function of Vivian's lively ideas.[126] They have no serious disputes; the dialectic in the dialogue is not primarily between the speakers but between Vivian's article and the dramatic frame Wilde called ''The Decay of Lying: An Observation.'' Through this dialectic of forms Wilde identifies with, and dissociates himself from, Vivian's doctrine: Expository prose is subsumed within artifice, and Wilde's critical discourse remains a fruitful lie, another form of ''art-criticism,'' which encloses but goes beyond the usual criticism-as-article.

Wilde's playful creativity is a model for the critic or artist as liar, who, as Vivian says, is ''the very basis of civilised society.'' When he is acknowledged as its leader,

> Art, breaking from the prison-house of realism, will run to greet him, and will kiss his false, beautiful lips, knowing that he alone is in possession of the great secret of all her manifestations, the secret that Truth is entirely and absolutely a matter of style; while Life—poor, probable, uninteresting human life—tired of repeating herself for the benefit of Mr. Herbert Spencer, scientific historians, and the compilers of statistics in general, will follow meekly after him, and try to reproduce, in her own simple and untutored way, some of the marvels of which he talks. (CW, 981)

The Con Artist as Critic: "Pen, Pencil and Poison"

There is more than a hint of the social subversive in this liar, and Wilde recognizes and emphasizes this subversive function in ''Pen, Pencil and Poison,'' an essay that first appeared in *The Fortnightly Review* (January 1889). In it Wilde and the poet, artist, and critic he describes, Thomas Griffiths Wainewright, invert and negate the role of respectable Victorian sage. Given Wilde's notions about the importance of suggestiveness (and even mystification) for both literature and criticism, it is no surprise that critics have interpreted this essay as an announcement of Wilde's new and more cynical persona, the artist as criminal.[127] We believe that Wilde's allusions and the critical questions he addresses show his dialectical interpretation and practice of Victorian cultural criticism. Adopting the form and the historical method of the ''Rise'' and ''The Truth of Masks,'' Wilde refashions and rescues from obscurity Wainewright's life, work, and ideas, offering them (much as he offered Greek historical criticism and Renaissance archeology) as a historical anticipation of his own modern Hellenism.

Since Wainewright was a man of letters and a criminal, his career provides an opportunity for Wilde to examine the relationships among morality, creativity, and criticism. Because he was a criminal, moralistic critics condemned his work and, according to Wilde, confused moral judgments with historical understanding and aesthetic evaluation. Using the form of a standard biographical, critical essay (like Arnold's on Shelley, Wordsworth, and Byron), Wilde singles out the qualities and beliefs which anticipate his own principles of aesthetic and cultural criticism, as well as certain personal attitudes and tastes. Wainewright, like Wilde, is said to have realized the "true harmony of all really beautiful things" (CW, 996), and that "all the arts were one" (CW, 1001). He, too, "tried to revive style as a conscious tradition" (CW, 998) and separated aesthetic from moral judgment. He believed, therefore, that taste or "right judgment" arose from an artistic temperament, and he developed a critical method based on the critic's ability to realize his own impressions of art, " 'to see the object as in itself it really is' " (CW, 997). Wainewright shared Wilde's taste for archeological accuracy in stage costume. Like Wilde, he appreciated the Romantic poets and disapproved of realistic and mimetic art. Wilde claims him, along with Ruskin and Browning, as a practitioner of "art-literature," which is the attempt to give "the literary equivalent for the imaginative and mental effect" of an artwork (CW, 998). Wilde notes with approval Wainewright's recognition that "Life itself is an art" (CW, 995), and that people can improve their artistic sense only by having " 'the best models constantly before their eyes' " (CW, 998).

Although Wilde's critique is basically positive, he adopts a historical perspective to recognize Wainewright's critical and imaginative limitations. He notes Wainewright's small and uneven creative oeuvre and his ahistorical method, which, Wilde explains, "did not belong to his day" (CW, 997). Wilde attacks Wainewright's critics from the same historical perspective. In criticism, he says, moral judgments are irrelevant to both historical understanding and aesthetic evaluation (CW, 1008). Moralizing critics and historians distort the past by intruding their moral sense "wherever it is not required" (CW, 1008). In keeping with his Hegelian, historicist position in the "Rise," Wilde invokes the "true historical sense" (CW, 1008) by refusing to equate artistic temperament and moral sense either in the subject of his memoir or in the practice of criticism: "There is no essential incongruity between crime and culture. We cannot re-write the whole of history for the purpose of gratifying our moral sense of what should be" (CW, 1008). Wilde, recognizing that murder must be condemned, does not deny the existence of a moral sense; instead, he demonstrates its prejudicial effect upon historical analysis and judgment.

Wilde appears to be disagreeing with a complex and interconnected body of literary and scientific thought when he offers Wainewright as an example of one who loved both nature and poetry (especially Wordsworth's poetry), and who was still an amoral criminal. Wordsworth's literary theory (based on Locke's and Hartley's associationist psychology) argued for realism in literature ("I have at all times endeavoured to look steadily at my subject") and the close relationship of aesthetic and moral judgments.[128] Wordsworth believed that moral character was formed and sustained by the benevolent influence of nature and poetry. Wilde disagreed. In the opening of "The Decay of Lying," when Vivian refuses to leave his books and "enjoy Nature" (CW, 970), he introduces an anti-Wordsworthian leitmotiv which includes his remarks that Wordsworth "found in stones the sermons he had already hidden there'' (CW, 977–78), that nature imitates art, and that nature is basically a bore rather than a ''great mother who has borne us'' (CW, 986).

In ''Pen, Pencil and Poison'' Wilde extended this attack on the necessary association of moral sense and creative or critical abilities: ''The fact of a man being a poisoner is nothing against his prose'' (CW, 1007). In so doing Wilde criticized not only the Wordsworthian view but also the critics who subscribed to it, notably W. C. Hazlitt, who edited Wainewright's *Essays and Criticism* (1880). Wilde contended, in response to Hazlitt, that ''the domestic virtues are not the true basis of art'' (CW, 1007) and that Wainewright's artistic temperament permitted him to enjoy the ''spiritual influences of Wordsworth's poetry'' (CW, 994) regardless of his criminal behavior.

Unquestionably, the portrait of the artist in ''Pen, Pencil and Poison'' is partially a projection of the person who penned it. Wilde would say that it could not have been otherwise: He was fascinated by Wainewright and by the issues of artistic and moral personality. As he remarked, ''One can fancy an intense personality being created out of sin'' (CW, 1007). Critics have interpreted a veiled confession of sin in the essay's subtitle (''A Study in Green'').[129] While such speculations may have more than a grain of truth in them, they distort a full understanding of Wilde's critical development by shifting attention to his personal life and by ignoring his literary and critical ideas. Wilde always toyed with confession and self-revelation in his life and his work, but the trial and the consequent revelation of his homosexuality have created a body of twentieth-century criticism which is, in its psychological obsession, analogous to the obsessive moralism of the Victorian criticism Wilde disliked. His interest, at least in literary and critical works, was in sin's relation to moralizing and creativity.[130]

In ''Pen, Pencil and Poison'' Wilde examined sin aesthetically. As he worked out the political and social implications of his synthesis in the 1880s, he developed ideas which appear in ''The Critic as Artist,'' ''A Chinese Sage,'' and ''The Soul of Man Under Socialism.'' He began to interpret sin from a Hegelian and evolutionary perspective as an ''essential element of progress'' (CW, 1023). Wainewright, understood historically, demonstrated social contradictions and so represented a force for change, but he had been branded a sinner antithetical to Philistine culture. Wilde has Wainewright explain to a visitor in jail the relation of his crime to the normal economic behavior of society: ''Sir, you City men enter on your speculations, and take the chances of them. Some of your speculations succeed, some fail. Mine happen to have failed, yours happen to have succeeded. That is the only difference, sir, between my visitor and me. But, sir, I will tell you one thing in which I have succeeded to the last. I have been determined through life to hold the position of a gentleman. I have always done so. I do so still'' (CW, 1006). Wainewright gains his leisure, and with it the opportunity for culture and creativity, by an act of criminal financial speculation: He kills his wives. While Wilde morally condemned Wainewright's actions, he recognized that they provided a solution, albeit unacceptable, to an important social contradiction for the middle-class artist (which Wilde himself had experienced): how to lead the life of a creative and cultured gentleman without inherited wealth.

The Critical Spirit as World Spirit in "The Critic as Artist"

Wainewright's solution was a creative but morally unsatisfactory example of the spirit of criticism manifesting itself as individual subversion or negation. In "The Critic as Artist" and "The Soul of Man Under Socialism," representing Wilde's most complex and fully articulated applications of his critical synthesis to problems of cultural transformation, he argued that all activities of mind, both creative and critical, are sinful or immoral from society's point of view because they are essential elements in the transformation of culture.[131] In "The Soul of Man" Wilde explains how socialism, operating in the practical realm, both embodies and encourages human spiritual development. In "The Critic as Artist" he describes how the critical spirit operates in the aesthetic realm, especially in regard to the creativity of artists and to the interpretation of both art and criticism. Together the two essays reinterpret the function of criticism and offer a utopian vision of culture as anarchy.

At the end of "The Critic as Artist" Gilbert reveals to Ernest the "True Function and Value of Criticism," which was the original title of this dialogue: "Creation is always behind the age. It is Criticism that leads us. The Critical Spirit and the World Spirit are one" (CW, 1058). They are one because, as Gilbert says elsewhere in the dialogue, "what is mind but motion in the intellectual sphere? The essence of thought, as the essence of life, is growth" (CW, 1048). Wilde is using Gilbert to restate the informing evolutionary idealist assumption of "The Decay of Lying," namely, that life uninformed by mind is dominated by an imitative instinct. In "The Critic as Artist" Wilde gives a new dialectical turn to his analysis of art and criticism. Among creative artists "it is the critical faculty that invents fresh forms. The tendency of creation is to repeat itself" (CW, 1021). Gilbert denies that the creative faculty is higher than the critical and suggests that "the antithesis between them is entirely arbitrary. Without the critical faculty, there is no artistic creation at all worthy of the name. You spoke a little while ago of that fine spirit of choice and delicate instinct of selection by which the artist realises life for us, and gives to it a momentary perfection. Well, that spirit of choice, that subtle tact of omission, is really the critical faculty in one of its most characteristic moods. . ." (CW, 1020). Wilde's dialectical analysis of the critic immediately breaks down the intellectual and social distinctions between creative and critical work, between subjective and objective truth. He asserts that a true critic can and should practice criticism as a creative art or as a formal or historical interpreter of texts (or as both simultaneously), and that his object will be to deepen the essential mystery and wonder of creative acts.

Wilde again builds his argument upon Hegelian philosophy and evolutionary theory. He assumes that the Hegelian ideal—absolute mind gradually realizing itself in history—can be achieved through the inheritance of acquired characteristics, that is, the hereditary transmission of culture. His idealist assumptions are clearest in his statements about the origins and the ends of life: "In every sphere of life Form is the beginning of things. The rhythmic harmonious gestures of dancing convey, Plato tells us, both rhythm and harmony into the mind. Forms are the food of faith, cried Newman. . . . He was right, though he may not have known how terribly right he was. The Creeds are believed, not because they are rational, but because they are repeated. Yes; Form is everything. It is the secret of life'' (CW, 1052).

Wilde also defines the individual and collective ends of life in idealist terms. Nature, he says, ''is matter struggling into mind'' (CW, 1040), a formulation which both Hegel and Herbert Spencer could accept. The end of life for individuals is to make ''instinct self-conscious and intelligent'' (CW, 1041), to develop the self-cultured, contemplative life endowed with *theoria,* the ''*ΒΙΟΣ ΘΕΩΡΗΤΙΚΟΣ*'' (*BIOS THEORETIKOS*), as ''the true ideal'' (CW, 1042)—thus the importance of what Wilde called ''doing nothing.''

Wilde again relies upon the theory of the inheritance of acquired characteristics to explain how instinct becomes self-conscious. As the notebooks show, he believed the individual's soul and imagination represent a collective inheritance (see CB, 145, 183). ''The imagination,'' says Gilbert, ''is the result of heredity. It is simply concentrated race-experience'' (CW, 1041). For this reason individual life is a kind of epiphenomenon: ''[I]t is not our own life that we live, but the lives of the dead, and the soul that dwells within us is no single spiritual entity, making us personal and individual, created for our service, and entering into us for our joy'' (CW, 1040–41).

To improve life the critical spirit, operating through creative and critical work, must transform the soul. In its creative mode, as art, it gives individuals the experience of harmony, of life expressed under the aspect of form, and so creates in the soul the ideal which is the impetus for improvement. It also provides a means to express and discover unrealized aspects of the soul, or to experience vicariously life in other times and cultures. In its utopian aspect it can provide new and better models that life, with its imitative instinct, can copy. As criticism it becomes a model of ''disinterested curiosity,'' making taste self-conscious and bringing the past alive through its use and development of the historical sense and through science. By living a life devoted to contemplation, the critic enacts the ideal of self-development. The critical spirit, in short, improves life by allowing us to see it as it is (and was), and as it is not (but might be).

In ''The Critic as Artist'' Wilde sets his utopian goals against an alternative and darker vision in which ''the scientific principle of Heredity,'' which has ''robbed energy of its freedom and activity of its choice'' (CW, 1040), simply sustains past and present social conditions. He sees the possibility, that is, of social progress or of decline, and he evaluates evolutionary change based on normative and pathological conceptions of the social organism derived from Aristotle, Buckle, Spencer, and Clifford (see CN, 30; CW, 1136). In the ''Rise'' Wilde offered Greece as the normal (i.e., rational) example of evolution of mind in the social organism because it was guided by the critical spirit. He says, however, that progressive social evolution is often the exception rather than the rule in history. Roman, Chinese, Hindu, and medieval European societies all stagnated and so developed pathologically. Hegel, therefore, underestimated the complexity of dialectical development in history (CW, 1106).[132]

Wilde distinguishes the situation of modern society by its problematic relationship with the critical spirit, which reemerged in the Renaissance and the Enlightenment. He argues that the modern development of mind (and so of progress) has met with an increasingly complex social organism dominated by an unthinking impulse toward stasis manifest in the institutions and customs society develops to survive. In the "Rise" Wilde describes this process as "the rational law lying at the base of all varieties of political changes [which is] that stability can result only from the statical equilibrium produced by the counteraction of opposing parts, since the more simple a constitution is the more it is insecure" (CW, 1129). Wilde also says that Herbert Spencer's theory of "The Instability of the Homogeneous," a general law common to all organic bodies, is an evolutionary interpretation of this theory (CW, 1130).

In "The Critic as Artist" Wildes bases his analysis of modern English society on the same theory of social development. He has Gilbert remark:

> The security of society lies in custom and unconscious instinct, and the basis of the stability of society, as a healthy organism, is the complete absence of any intelligence amongst its members. The great majority of people being fully aware of this, rank themselves naturally on the side of that splendid system that elevates them to the dignity of machines, and rage so wildly against the intrusion of the intellectual faculty into any question that concerns life, that one is tempted to define man as a rational animal who always loses his temper when he is called upon to act in accordance with the dictates of reason. (CW, 1044)

His description of social problems, presented in these evolutionary terms, also includes a synthesis of Arnold's disparagement of Philistinism and J. S. Mill's fear of social stagnation (expressed most memorably in *On Liberty,* the only book of Mill's that Wilde admired).

Wilde sees Puritanism in its moral, economic, and artistic manifestations as a static force resisting the critical spirit: Its unconscious assumptions and its resistance to self-examination produce mass stupidity for the sake of social stability. This belief, which appears in the notebooks, is basic to Wilde's social criticism in "The Critic as Artist."[133] Near the beginning of the essay Ernest and Gilbert agree that the English public "feels perfectly at its ease when a mediocrity is talking to it" and that it "forgives everything except genius" (CW, 1009). They observe that journalism, which "justifies its own existence by the great Darwinian principle of the survival of the vulgarest" (CW, 1015), is the institution that the English have invented to propagate "the ignorance of the community" (CW, 1048).

As Wilde also argued in "Pen, Pencil and Poison" (CW, 1008), this ignorance has created a confusion in which behavior appropriate to the practical, "lower and less intellectual" realm of ethics is prescribed for the unpractical, higher realm of art and science, beauty and truth (CW, 1048). As a result, English "ideals are emotional and not intellectual" (CW, 1044), and "humanitarian" solutions to English social problems are shallow, immediately practical, finally self-interested, and consequently tools to perpetuate rather than solve the problems they address. Summing up Wilde's analysis and criticism of Philistine culture, Gilbert remarks, "It takes a thoroughly selfish age, like our own, to deify self-sacrifice. It takes a thoroughly grasping age, such as that in which we live, to set above the fine intellectual virtues, those shallow and emotional virtues that are an immediate practical benefit to itself. They miss their aim, too, these philanthropists and sentimentalists of our day, who are always chattering to one about one's duty to one's neighbour"(CW, 1043). Instead of enriching culture, England has "done one thing; it has invented and established Public Opinion, which is an attempt to organize the ignorance of the community, and to elevate it to the dignity of physical force. But Wisdom has always been hidden from it. Considered as an instrument of thought, the English mind is coarse and undeveloped. The only thing that can purify it is the growth of the critical instinct" (CW, 1055–56). Wilde sees modern England, dominated by materialism in philosophy and practical concerns in life, as a highly developed pathological state whose "health" (i.e., social stability) represses the further development of mind and so becomes a thesis in Wilde's Hegelian historical dialectic.

Gilbert relies on an idealist evolutionary theory of mind and language (Müller's?) to condemn the idea that progress can be achieved by action: "There is no mode of action, no form of emotion, that we do not share with the lower animals. It is only by language that we rise above them, or above each other—by language which is the parent, and not the child, of thought" (CW, 1023). Because it is necessarily uninformed by imagination and critical instinct, action is the lowest form of human activity. As Gilbert says, action "is a blind thing dependent on external influences, and moved by an impulse of whose nature it is unconscious. It is a thing incomplete in its essence, because limited by accident, and ignorant of its direction, being always at variance with its aim. Its basis is the lack of imagination. It is the last resource of those who know not how to dream" (CW, 1023). In Wilde's theory the purpose of action is survival, and those who exist, who do and act uncritically, simply reproduce nature's order as tools of its drive for stability. However, those who "sin" by contemplating and resisting the demands of action actually advance critical consciousness and encourage change. This is "the importance of doing nothing," the subtitle of the first half of the dialogue.

Wilde, relying upon Aristotelian, Hegelian, and modern scientific theories, provides his cure for the body politic, an antithetical redefinition of health as change, the expression of energy and the spirit coming to fuller awareness of itself.[134] His model for health is, as usual, the ancient Greeks, "the nation of art-critics" (CW, 1044), who, unlike the English, exercised the critical spirit in all aspects of their civilization (CW, 1016). This spirit, the desire for and love of harmony, is realized in only a few moderns. In its developed state, as contemplation and the critical instinct, it creates health as wisdom in the form of great art, utopian thought, and scientific achievement; for this reason, "in the opinion of society, Contemplation is the gravest sin of which any citizen can be guilty, [while] in the opinion of the highest culture it is the proper occupation of man" (CW, 1039).

To Plato and Aristotle, Gilbert says, contemplation was the passion for wisdom and knowledge, "the noblest form of energy" (CW, 1039). For Wilde contemplation as a process serves the same function as the academy served for Matthew Arnold. Like Arnold, Wilde believed that "the stress and turmoil of actual existence" was an environment which "degraded [Thought] by its constant association with practice." That is, in a society motivated by practical self-interest, who,

Gilbert asks, "can seriously claim to be able to form a disinterested intellectual judgment about any one thing?" (CW, 1042). Arnold imagined an institutional best and disinterested self, but Wilde relied upon the critical instinct and contemplative energy. As Gilbert asserts, "By revealing to us the absolute mechanism of all action, and so freeing us from the self-imposed and trammelling burden of moral responsibility, the scientific principle of Heredity, has become, as it were, the warrant for the contemplative life. It has shown us that we are never less free than when we try to act" (CW, 1040). Only the development of mind through contemplation will improve society by yielding disinterested truths or imaginative creativity.

In "The Critic as Artist" Wilde's critic differs significantly from the public educator of "The English Renaissance," who dispenses information and encourages respect for art and artists. The later essay proposes a larger and more complex function: The model critic should rely on critical and creative contemplation to produce work which will be perceived as sin because it opposes the stability of the dominant culture. "What is termed Sin is an essential element of progress. Without it the world would stagnate, or grow old, or become colourless. By its curiosity Sin increases the experience of the race. Through its intensified assertion of individualism it saves us from monotony of type. In its rejection of the current notions about morality, it is one with the higher ethics" (CW, 1023–24).

Wilde recognizes cultural contradictions as necessary and progressive. The assertion of individualism, the substitution of sinful antitheses for stagnant received truths, the overturning of traditional goals for critics and criticism—all function to renew and develop the cultural self-awareness of society. Wilde, then, offers not a collection of critical insights but a model of critical and creative process for the imitative instinct, which he hopes will turn imitation against itself. As Gilbert puts it, "The influence of the critic will be the mere fact of his own existence. He will represent the flawless type. In him the culture of the century will see itself realised" (CW, 1053).

This modeling of process is crucial to Wilde's theory of cultural improvement precisely because, as the notebooks show, he believed with Spencer that evolutionary progress was toward differentiation and specialization of function, toward more freedom and more individualism (i.e., toward a society based on voluntary cooperation and mutual aid; see CB, 7–9, 14, 69). The educational corollary is Wilde's idealist "self-culture" based on *theoria,* a process of self-realization: "The Greeks did this, and when we come in contact with the Greek critical intellect, we cannot but be conscious that, while our subject-matter is in every respect larger and more varied than theirs, theirs is the only method by which this subject-matter can be interpreted" (CW, 1055). That nation of art critics has "left us, as their legacy to modern thought, the conception of the contemplative life as well as the critical method by which alone can that life be truly realised" (CW, 1044). The contemplative life, as Wilde described it, provided both the goal and methods of true education. In his 1881 poem "Theoretikos" and in this essay Wilde invokes contemplation as both an escape from modern life and a means of improving it.[135] Gilbert remarks, "Is such a mode of life unpractical? Ah! it is not so easy to be unpractical as the ignorant Philistine imagines. It were well for England if it were so. There is no country in the world so much in need of unpractical people as this country of ours. . . . What we want are unpractical people who see beyond the moment, and think beyond the day. . . . It is through the voice of one crying in the wilderness that the ways of the gods must be prepared.''[136]

Throughout the dialogue Wilde plays variations upon the themes of process and change, of the liberation and expression of spiritual energy. A study of the artistic process leads to a dialectical understanding of the artist as critic as well as the critic as artist. Ernest denies this relation early in the dialogue when he asserts that artists, endowed with a spirit of choice and an instinct for selection, have no need for criticism (CW, 1011). Gilbert, Wilde's theoretical voice throughout, asserts the necessary relation. Echoing Ernest, he argues that the creative artist practices criticism through his ''fine spirit of choice and delicate instinct of selection,'' as well as by a ''subtle tact of omission.'' In this sense, Gilbert says, ''all fine, imaginative work'' and all great artists are ''self-conscious and deliberate'' (CW, 1011, 1020). He broadens his attack on Ernest's anti-intellectual theory by explaining that all creative ages in art are also critical ages: ''It is to the critical instinct that we owe each new school that springs up, each new mould that art finds ready to its hand'' (CW, 1021). With Ernest as foil, Gilbert argues that critical practice ''outside creation'' is higher and more difficult than creativity because talking and thinking, rather than doing, are the essential human activities (CW, 1022). On this principle Gilbert establishes the following hierarchy of activities: Just as art criticism (talking about art) is a higher activity than art (doing or making creative things) so art (a kind of talking about life) is a higher activity than a life of action in the world.

In Hegelian fashion Gilbert concludes that the mind's ability to know itself is the highest human value. It follows that the highest criticism is the most self-absorbed, the freest from the influence of the external world and from judgments by its standards. Art, which takes existence for its raw materials and is often judged by standards of verisimilitude, could never be the purest expression of mind. Only the highest criticism, ''the record of one's own soul,'' has ''least reference to any standard external to itself, and is, in fact, its own reason for existing, and, as the Greeks would put it, in itself, and to itself, an end'' (CW, 1027). Because Wilde believes that the human essence, mind and spirit, should be developed in the world, he exalts the dreamer, who contemplates and criticizes, over the man of action. The highest form of criticism is subjective and creative, the purest expression of the soul's impression; it is, in the much-excerpted phrase, ''the only civilised form of autobiography'' (CW, 1027). The critic uses art as raw material for his creativity, much as the artist uses form, color, and the unseen world of passion and thought. That is, to be creative the critic should ''see the object as in itself it really is not'' (CW, 1030).

Wilde's Hegelian logic is impeccable: The best realization of critical spirit comes when the critic is least mimetic. The more subjective the expression of a critic-as-artist, the more objective is the manifestation of the World-Spirit. Subjectivity becomes the portal for discovery of Hegelian reality. Again, metaphysics liberates Wilde from Pater's materialist version of subjectivity. Seeing the object ''as in itself it really is not'' offers a momentary and incomplete revelation of the World-

Spirit: The subjective becomes objective, and the individual transcends the limitations of materialist impressionism by giving objective form to that Spirit. Gilbert's description of the critic as creator focuses upon a dialectical process:

> To the critic the work of art is simply a suggestion for a new work of his own, that need not necessarily bear any obvious resemblance to the thing it criticises. The one characteristic of a beautiful form is that one can put into it whatever one wishes, and see in it whatever one chooses to see; and the Beauty, that gives to creation its universal and aesthetic element, makes the critic creator in his turn, and whispers of a thousand different things which were not present in the mind of him who carved the statue or painted the panel or graved the gem. (CW, 1030)

Furthermore, artists using the critical spirit, the sense of harmony, create works of art which allow the critic—or any perceiver—to see in them "whatever one chooses to see" and so to express and create beauty of their own. Wilde's critic manifests creativity imaginatively by re-seeing and reinterpreting the beautiful, while beauty, by its very nature, encourages such interpretation. This creative criticism constitutes a more developed consciousness than art because it "criticises not merely the individual work of art, but Beauty itself" (CW, 1030), and offers creative artists fresh forms and ideas for new creation. Ending the discussion before supper, Gilbert describes a dialectical development through which beauty encourages and liberates further expressions of beauty.[137]

Although Wilde's primary concern is the critic, he also has Gilbert and Ernest discuss the psychology of response to art. Like all theories which assume the mind's shaping power, Wilde's idealist, racialist theory asserted that individual minds inevitably produced variant interpretations of artworks. He valued such variations, seeing them as expressions of personality and creativity rather than as undesirable aberrations. He found in Aristotle's concern with catharsis a Hellenic origin and warrant for his interests in psychology. Catharsis is proof of his general belief that the expression of energy is healthy and the suppression of it pathological: "To have a capacity for a passion and not to realise it, is to make oneself incomplete and limited" (CW, 1018). Tragedy produces emotions which purify and spiritualize the beholder. "[N]ot merely does it spiritualise him, but it initiates him also into noble feelings of which he might else have known nothing" (CW, 1018). Wilde believed that the "true and only meaning" of catharsis might be analogous to an initiation rite, a kind of aesthetic liberation or awakening of the human soul. He interpreted this Aristotelian pragmatics of response as Platonic theory materialized: "[Plato] first perhaps stirred in the soul of man that desire that we have not yet satisfied, the desire to know the connection between Beauty and Truth, and the place of Beauty in the moral and intellectual order of the Kosmos. The problems of idealism and realism, as he sets them forth, may seem to many to be somewhat barren of result in the metaphysical sphere of abstract being in which he places them, but transfer them to the sphere of art, and you will find that they are still vital and full of meaning" (CW, 1018). We see this relation between Aristotle's and Plato's aesthetic theories as a Hellenic parallel to Wilde's Hegelian theory: Wilde revitalized Hegelian logic and analysis by applying them to the concrete realms of art and life rather than to the metaphysical realm of formal philosophy.

Part II of the dialogue opens by considering the critic as "interpreter," one who analyzes art by seeing the object as in itself it is. The interpretive critic can begin, Gilbert says, with "an analysis or exposition of the work itself" (CW, 1032), for which he can provide a scholarly and historical context. "To understand Shakespeare truly" one needs a synthesis of various histories, including ideas of the time, theatrical and linguistic history, the dialectical history of forms, and, importantly in Wilde's historical aesthetics, one must grasp the relationship of Shakespeare to Greek drama: "In a word, he must be able to bind Elizabethan London to the Athens of Pericles, and to learn Shakespeare's true position in the history of European drama and the drama of the world" (CW, 1033). Interpretive criticism, however, should not provide reductive explanations but rather deepen the "mystery" or "wonder" of the artwork and its maker. This goal is characteristic of nineteenth-century critics like Arnold, who thought that art would replace religion as a consolation and an expression of ultimate mystery. For Wilde, of course, art was the sensuous embodiment of the Spirit, its consolation was beauty, and its revelation was the soul of each and every person.

In Gilbert's system the interpretive critic is not a self-effacing worshiper but rather one who makes interpretation stronger and truer by intensifying personality. "Personality," he says, "is an absolute essential for any real interpretation" (CW, 1034). Like Ruskin he sees true interpretive ability as a rare gift, and right interpretation, like art itself, as a mode of revelation which requires a dialectical relation of enlightened and cultured personalities: "[A]s art springs from personality, so it is only to personality that it can be revealed, and from the meeting of the two comes right interpretive criticism" (CW, 1034).

Just as creative critics express and promote the innovative instinct, so, too, do interpretive critics. Both will increasingly shift attention from life and its "sordid security" to art (CW, 1035). Existence, Gilbert says, causes pain and suffering and does not allow individuals the freedom to experience perfection or to perfectly express or understand their personalities. However, in the contemplation of art "there is no passion that we cannot feel, no pleasure that we may not gratify, and we can choose the time of our initiation and the time of our freedom also. . . . It is through Art, and through Art only, that we can realise our perfection" (CW, 1038). So, Wilde returns via the creative and interpretive functions of criticism to one of his assumptions: Art is at once the expression of the artist's creativity, the focus which provokes the critic's creativity, and the medium through which all "elect spirits" (CW, 1034) realize themselves through contemplation.

Wilde's idea that art, rather than metaphysics or religion, is the only satisfactory form of understanding and expression for the modern critical and cultured person recapitulates his earlier disagreements with Hegel's theory of the increasing spiritualization of art and culture. In the Commonplace Book Wilde had first seen the problem of the inadequacy of philosophy applied to modern life; in a note recording a remark of Hegel's he wrote, "If Philosophy aims at doing good to man it comes too late in the day . . ." (CB, 189). When he returned to the idea in "The Critic as Artist," he explained that religion, too, was out of date; our fathers, he says, "have exhausted the faith faculty of the species. . . . We cannot go back to the philosopher, and the mystic leads us astray" (CW, 1040). In

Wilde's synthetic system there was a necessity for art, mediated by criticism, to fulfill the functions of philosophy and religion. He offers such a possibility through an analogy: "[J]ust as Nature is matter struggling into mind, so Art is mind expressing itself under the conditions of matter, and thus, even in the lowliest of her manifestations, she speaks to both sense and soul alike" (CW, 1040). Art combines both the subjective and objective spheres, the realms of mind and matter.

The critic, because he lives primarily through contemplation, provides the model for the future: "[A]s civilisation progresses and we become more highly organised, the elect spirits of each age, the critical and cultured spirits, will grow less and less interested in actual life, and *will seek to gain their impressions almost entirely from what Art has touched*" [emphasis in original] (CW, 1034). Through immersion in the experience and practice of art and scholarship the critic achieves knowledge of himself and of past and present cultures; in the process he perfects the culture of the future. This can happen, Gilbert and Wilde believe, because culture and imagination are concentrated and collective racial experience. The uneducated individual is unconscious of this culture. Contemplation of art makes the unconscious self-conscious: "To realise the nineteenth century, one must realise every century that has preceded it and that has contributed to its making. To know anything about oneself one must know all about others" (CW, 1040). The critic also improves culture by developing his own knowledge, taste, and reason, by practicing his art upon the received culture:

> The culture that this transmission of racial experiences makes possible can be made perfect by the critical spirit alone, and indeed may be said to be one with it. For who is the true critic but he who bears within himself the dreams, and ideas, and feelings of myriad generations, and to whom no form of thought is alien, no emotional impulse obscure? And who the true man of culture, if not he who by fine scholarship and fastidious rejection has made instinct self-conscious and intelligent, and can separate the work that has distinction from the work that has it not, and so by contact and comparison makes himself master of the secrets of style and school, and understands their meanings, and listens to their voices, and develop that spirit of disinterested curiosity which is the real root, as it is the real flower, of the intellectual life, and thus attains to intellectual clarity, and, having learned "the best that is known and thought in the world," lives—it is not fanciful to say so—with those who are the Immortals. (CW, 1041)

Here is the critic—the fittest in thought—practicing the self-conscious, selective process which ensures the survival of the fittest in thought.[138]

Not only is he a "flawless type" for imitation, but the critic as prophet, inspired by contemplation, has influential cultural power. As Gilbert argues, the critic, acting as influence, "will concern himself not with the individual, but with the age, which he will seek to wake into consciousness, and to make responsive, creating in it new desires and appetites, and lending it his larger vision and his nobler moods" (CW, 1053). Wilde describes such critics as those who, through self-culture of inherited temperament, develop taste, critical spirit, and a sense of beauty:

> Temperament is the primary requisite for the critic—a temperament exquisitely susceptible to beauty, and to the various impressions that beauty gives us. Under

> what conditions, and by what means, this temperament is engendered in race or individual, we will not discuss at present. It is sufficient to note that it exists, and that there is in us a beauty-sense, separate from the other senses and above them, separate from the reason and of nobler import, separate from the soul and of equal value—a sense that leads some to create, and others, the finer spirits as I think, to contemplate merely. (CW, 1049)

The elevation of critical temperament over creative imagination has seemed to later critics a kind of self-serving rhetorical inversion, but it is a logical and necessary consequence of Wilde's theory. Because he understands social growth rather than stasis as the norm of modern society, temperament and the independent analytic ability it engenders are crucial prerequisites for the critic: "[T]he subject-matter at the disposal of creation is always diminishing, while the subject-matter of criticism is increasing daily. There are always new attitudes for the mind, and new points of view. The duty of imposing form upon chaos does not grow less as the world advances. There was never a time when Criticism was more needed than it is now. It is only by its means that Humanity can become conscious of the point at which it has arrived" (CW, 1055).

Preparing for his conclusion, Wilde has Gilbert enumerate the social, political, and institutional uses of criticism: It will supplant education by memorization and thus make the mind a fine instrument; its concentration of intellect upon the mass of creative work literally will make culture possible by canonizing those texts which are the thread through the labyrinth of past work; in its philosophical and archeological modes it will uncover the unwritten history of centuries and will prove to be "the exact science of mind in the process of becoming" (CW, 1056); it will teach a cosmopolitan worldview inclined toward the rightful use of power in pursuit of peace; and it will help annihilate racial prejudice. Going beyond Arnold's confederation in "The Function of Criticism," Wilde imagines a European utopia united by criticism: "Intellectual criticism will bind Europe together in bonds far closer than those that can be forged by shopman or sentimentalist. It will give us the peace that springs from understanding" (CW, 1057).

Gilbert cites Darwin's "sweet reasonableness" to exemplify the last and most important attribute of the critical temper. Criticism, says Gilbert, "recognizing no position as final, and refusing to bind itself by the shallow shibboleths of any sect or school, creates that serene philosophic temper which loves truth for its own sake, and loves it not the less because it knows it to be unattainable" (CW, 1057). As an antinomian spokesman for mind, Gilbert concludes by inverting the prejudice of the English middle class, that intelligence is sinful: "There is no sin except stupidity" (CW, 1057).

In his utopia Gilbert imagines a "true culture" in which "sin is impossible" (CW, 1058), implying a transformation, not a rejection, of ethics. Earlier in the dialogue Gilbert commented that "the mere existence of conscience, that faculty of which people prate so much nowadays, and are so ignorantly proud, is a sign of our imperfect development. It must be merged in instinct before we become fine" (CW, 1024). After instinctive conscience has perfected individuals, they "can do everything they wish without hurt to the soul, and can wish for nothing that can do the soul harm" (CW, 1058). Contemporary English ethics, however, are based on "a vulgar standard of goodness" which "requires a certain amount of sordid terror, a certain lack of imaginative thought, and a certain low passion for middle-class respectability" (CW, 1057–58). To transform this culture, Gilbert asserts that "AEsthetics are higher than ethics" and provides the master analogy drawn from Wilde's synthesis of idealism and evolutionary theory: "AEsthetics, in fact, are to Ethics in the sphere of conscious civilisation, what, in the sphere of the external world, sexual is to natural selection. Ethics, like natural selection, make existence possible. AEsthetics, like sexual selection, make life lovely and wonderful, fill it with new forms, and give it progress, and variety and change" (CW, 1058).

This analogy represents Wilde's use of Darwin's theory in *The Descent of Man* concerning the roles of natural selection and sexual selection in "the advancement of the welfare of mankind."[139] Darwin concluded that "man is descended from some lowly organized form," has "risen . . . to the very summit of the organic scale," and may have "hope for a still higher destiny in the distant future" (DM, 707). However, the means for mankind's advancement at present remained for Darwin an "intricate problem" (DM, 706). Man had "advanced to his present high condition through a struggle for existence consequent on his rapid multiplication," that is, through natural selection. Darwin believed that such a "severe struggle" should still continue "if he is to advance still higher." On the other hand, he recognized that "other agencies," such as "habit, the reasoning powers, instruction, religion, etc.," are more important than natural selection to develop "the highest part of man's nature" (DM, 706–7). In fact, as Darwin's full title—*The Descent of Man and Selection in Relation to Sex*—implies, he saw sexual selection as necessarily important in evolution:

> He who admits the principle of sexual selection will be led to the remarkable conclusion that the nervous system not only regulates most of the existing functions of the body, but has indirectly influenced the progressive development of various bodily structures and of certain mental qualities. Courage, pugnacity, perseverance, strength and size of body, weapons of all kinds, musical organs, both vocal and instrumental, bright colors and ornamental appendages, have all been indirectly gained by the one sex or the other, through the exertion of choice, the influence of love and jealousy and the appreciation of the beautiful in sound, color or form; and these powers of the mind manifestly depend on the development of the brain. (DM, 705–6)

Human progress for Darwin was a complex problem which required a complex solution. In society he continued to favor natural selection, "open competition for all men; and the most able should not be prevented by laws or customs from succeeding best and rearing the largest number of offspring" (DM, 706–7). Darwin also thought man might "do something not only for the bodily constitution and frame of his offspring, but for their intellectual and moral qualities" through sexual selection (DM, 706). In nature, Darwin suggested, sexual selection operated through two agencies, male struggle and female choice. In the first, males struggle to drive away or kill their rivals so that they can mate with passive females; in the second, members of the same sex (generally males) struggle among themselves to "excite or charm those of the opposite sex, generally the females, which no longer remain passive, but select the more agreeable partners" (DM, 702). Darwin saw sexual selection by female choice in nature as analogous to artificial selection practiced by man in the breeding of domestic animals. Further, he saw the process as extending to man's conscious choice in the selection of a marriage partner.[140]

Wilde's analogy of ethics and aesthetics to natural and sexual selection shows the influence of the Darwinian notion of sexual selection on his work and his differences with Darwin on the causes of social progress.[141] In Wilde's analogy only sexual selection was progressive, while natural selection and the ethics which it generated simply continued to make existence possible. They were a stabilizing, inertial force, like the artistic realism he decried in "The Decay of Lying." Therefore, Wilde believed it was time to invert the previous order, to emphasize aesthetics rather than ethics in conscious civilization and sexual rather than natural selection in the external world. Of course nature had good intentions, but they could be realized only when criticism and the critical spirit operated in society. So, Gilbert concludes, "The nineteenth century is a turning point in history, simply on account of the work of two men, Darwin and Renan, the one the critic of the Book of Nature, the other the critic of the books of God. Not to recognise this is to miss the meaning of one of the most important eras in the progress of the world. Creation is always behind the age. It is Criticism that leads us. The Critical Spirit and the World Spirit are one" (CW, 1058). "The Critic as Artist," far from being a decadent document of the art-for-art's-sake movement, is a manifesto stating Wilde's humanist cultural criticism and his program for utopian social and personal development.

Anarchy and Culture: The Evolutionary Turn of Wilde's Cultural Criticism[142]

In *Culture and Society, 1780–1950*, Raymond Williams defines a tradition of cultural criticism which began with Burke and Cobbett and remains a living force today. In that tradition Williams sees Wilde as a figure in an "Interregnum" between 1880 and 1914. He provides "a brief, separate section" about the writers in the interregnum because, he argues, "if they were neglected altogether, certain important links would be missing. Yet we shall not find in them, except perhaps in Hulme, anything very new."[143] Though Williams describes Oscar Wilde "rather than Pater [as] the first of the minor inheritors of Arnold" (C & S, 172), we disagree with his judgment that Wilde added little that was new to the Arnoldian tradition of cultural criticism.

We define Wilde's place in English cultural criticism by contrasting his and Arnold's application of ethnological and evolutionary theories to cultural theory. We argue that while both men attacked the perversions of Philistine culture and stressed the importance of curiosity and the critical spirit as the proper guides to cultural improvement,[144] they disagreed fundamentally over the nature of man and the ways in which personal and social improvement could and should occur.

Arnold's position in *On the Study of Celtic Literature* and *Culture and Anarchy* is contradictory. In both he states that ethnology accounts for cultural characteristics, that is, that qualities which we now know to be learned are biologically determined by race. At the same time, in both essays he argues that these cultural characteristics can and should be changed through educational institutions.

Wilde's cultural criticism demonstrates a more logical application of ethnological and evolutionary theory to methods of cultural improvement. Beginning with a belief that aesthetic and creative abilities are racially inherited, Wilde develops (principally in "The Soul of Man Under Socialism") a scheme for both the individual and social development of these characteristics. He envisions a nonauthoritarian socialist state which would free individuals to pursue true self-realization through autodidactic education. Such an anarcho-communist state would also end the competitive struggle for existence in human society and would substitute sexual selection (eugenics) as the means of racial, and consequently cultural, improvement. He therefore recognized the connections among cultural institutions and the consequent need for reforms not only in education but in all those social institutions which profoundly affected human development.

Race and Education: The Contradictions of Arnoldian Culture

The importance of ethnology in Arnold's cultural criticism is most evident in *Culture and Anarchy* and *On the Study of Celtic Literature*. Despite the fact that several critics—most notably Frederic Faverty—have extensively described Arnold's knowledge and use of ethnology and philology, most readers would probably still agree with Northrop Frye's comment that "culture for Arnold is a whole of which the church forms part, but as culture is not, like church, the name of a specific community, the problem of defining the community of spiritual authority is still with us."[145] This view is not altogether accurate. Arnold found in racialist theory a "scientific" definition for his community of spiritual authority,[146] and on this definition he based, perhaps paradoxically, his educational theory of cultural improvement.

In critical accounts of *Culture and Anarchy* Arnold's dependence on the contemporary science of ethnology, which underlies the entire critique and concludes the fourth chapter, "Hebraism and Hellenism," is so often neglected that it should be quoted: "Science has now made visible to everybody the great and pregnant elements of difference which lie in race, and in how signal a manner they make the genius and history of an Indo-European people vary from those of a Semitic people. Hellenism is of Indo-European growth, Hebraism is of Semitic growth; and we English, a nation of Indo-European stock, seem to belong naturally to the movement of Hellenism."[147] *On the Study of Celtic Literature* contains, as Frederic Faverty notes, "the most explicit and fullest expression"[148] of Arnold's racial theory. As Faverty explains, Arnold relied upon the racialist philological studies of Renan, Max Müller, and others for his basic idea of the Indo-European migration and the dispersion of racial strains into western Europe. He believed that English national character was explainable as a consequence of the historical intermixture of three Indo-European strains: the Celtic (which included the Irish and

French), the Norman (which contained the Latin strain), and the Teutonic (which included the Saxon and German). The centrality of this racial theory to Arnold's cultural analysis is apparent from the following passage in the Celtic lectures: "Modes of life, institutions, government, climate, and so forth,—let me say it once for all,—will further or hinder the development of an aptitude, but they will not by themselves create the aptitude or explain it. On the other hand, a people's habit and complexion of nature go far to determine its modes of life, institutions, and government, and even to prescribe the limits within which the influences of climate shall tell upon it."[149]

The rationale for racial influence on national culture led Arnold to believe that a nation's consciousness of its racial heritage could radically change its cultural life.[150] In both *Culture and Anarchy* and *On the Study of Celtic Literature* he argued that a false conception of racial heritage had distorted the understanding, and therefore the development, of the English national character. In *Culture and Anarchy* Arnold said that the dominance of Hebraism in English culture had been fostered by the middle-class Anglo-Saxon's mistaken belief in his racial kinship with the Semites, whereas "the actual instincts and forces" (C & A, 175), the innate genius, of the English people were Indo-European, and belonged in the Hellenic mainstream of European history.[151] At the conclusion of *On the Study of Celtic Literature* Arnold proposed to "reunite ourselves with our better mind and with the world through science . . . [by founding] at Oxford a chair of Celtic" (OSCL, 386). Pursuing an ethnological analysis similar to that in *Culture and Anarchy*, Arnold argued that "so long as we are blindly and ignorantly rolled about by the forces of our nature, their contradiction baffles us and lames us; so soon as we have clearly discerned what they are, and begun to apply to them a law of measure, control, and guidance, they may be made to work for our good and to carry us forward" (OSCL, 383).

Arnold believed that education provided the best hopes for social and cultural improvement. To raise the racial consciousness of the English and so remove the artificial impediment which restrained the power of their national "genius," Arnold proposed a centrally organized system of national education. A small group of "sovereign educators" (C & A, 229) or "*aliens*" (C & A, 146) who embody the best self would correct two major manifestations of ignorance and blindness, the mistaken beliefs that the English were racially related to the Hebrews and that they were not related to the Celts.

It may seem incongruous that Arnold, believing in the humanizing power of culture, would turn to nature for his educators. However, in the light of his racial theory this choice makes perfect sense. In *Culture and Anarchy* he said that the saving remnant of individuals who might constitute the cultural authority missing in English life was indeed a *natural* elite: "But in each class there are born a certain number of natures with a curiosity about their best self, with a bent for seeing things as they are, for disentangling themselves from machinery, for simply concerning themselves with reason and the will of God, and doing their best to make these prevail;—for the pursuit, in a word, of perfection" (C & A, 145). A passage which shortly follows these comments shows that Arnold literally meant that a certain number of these "*alien*" natures are "born": "I mean, the number of those who will succeed in developing this happy instinct will be greater or smaller, in proportion both to the force of the original instinct within them, and to the hindrance or encouragement which it meets from without'' (C & A, 146).

Arnold recognized that nature's cultural elite already hindered "the unchecked predominance of that class-life'' (C & A, 146). But, as he reiterated at the conclusion of the last three chapters of *Culture and Anarchy*, the dominance of class consciousness and the confusion it caused would not be ended until these "sovereign educators'' were organized as cultural authorities. They would recognize that "master-impulse even now of the life of our nation,'' which would reveal to England its natural path of perfection, the "firm intelligible law of things'' governing the development of its Hellenic and Indo-European genius (C & A, 229).

On the Study of Celtic Literature provides a practical instance of Arnold's educational theory. There he argued that the insolence, hardness, lack of intelligence, and fumbling which characterized the Philistine domination of Ireland would be changed if the English recognized their racial relationship to the Celts. To end centuries of Anglo-Irish strife and to liberate imaginative reason, Arnold proposed the establishment of a chair of Celtic Studies at Oxford. In arguing that the saving remnant of "sovereign educators'' could improve society by canonizing Celtic culture, Arnold assumed that a change in curricular content would provide the individual student with the necessary knowledge to choose between his Celtic, or better, self and his dominant Philistinism. Ultimately, for Arnold, only the individual's personal enlightenment and moral and conscious choice should be the effective agencies of beneficial social change.

Radical Darwinism, "A Chinese Sage," and "The Soul of Man Under Socialism"

Although Arnold thought that the recognition of the Celtic strain in English character would improve culture, he was not sanguine about the political liberation and recognition of Ireland.[152] This political position probably led him to abandon the ethnological authority of a man who had considerable influence on his thinking: "M. Renan has a theory that according to 'that great law by which the primitive race of an invaded country always ends by getting the upper hand, England is becoming every day more Celtic and less Germanic.' ''[153] The theory of the eventual political triumph of Celts over English, which Arnold did not accept, was still of cultural and political importance more than twenty years later when Grant Allen—anthropologist, man of letters, and supporter of Irish nationalism—published in *The Fortnightly Review* "The Celt in English Art,'' in which he described "the return wave of Celtic influence over Teutonic or Teutonized England'':[154]

> The great and victorious aesthetic movement . . . may be regarded in its wider aspect as just a particular part of the general racial, political, and social return-wave. It is a direct result, I believe, of the Celtic reflux on Teutonic Britain, and of the resurgence of the Celtic substratum against Teutonic dominance. . . . The Celt comes back upon us with all the Celtic gifts and all the Celtic ideals—

> imagination, fancy, decorative skill, artistic handicraft; free land, free speech, human equality, human brotherhood. How significant the fact that Henry George bears a Cornish surname, and that Alfred Russell Wallace (whose very patronymic means the Waleys or Welshman) was born at Usk, in purely Cymric Monmouthshire! (CEA, 272)

Allen's racialist analysis affirmed Renan's theory that the conquered Celts would end by getting the upper hand in England. He described a broad range of individuals, groups, ideas, and issues, all of which he saw as manifestations of this Celtic reflux. To emphasize the "connexion between the decorative revival and the Celtic upheaval of radicalism and socialism," Allen pointed specifically to William Morris and "Mr. Oscar Wilde, whom only fools ever mistook for a mere charlatan, and whom wise men know for a man of rare insight and strong common-sense" (CEA, 273).

Wilde read Allen's essay, which appeared in the same issue of *The Fortnightly Review* as "The Soul of Man Under Socialism," and immediately recognized an intellectual and political ally. In a letter to him written during the week of the essay's publication, Wilde said: "I beg you will allow me to express to you my real delight in your article in the *Fortnightly*, with its superb assertion of that Celtic spirit in Art that Arnold divined, but did not demonstrate, at any rate in the sense of scientific demonstration, such as yours is."[155] In "The Critic as Artist" Wilde shows that he shared Allen's belief in the racial origins of the aesthetic movement: "And though the mission of the aesthetic movement is to lure people to contemplate, not to lead them to create, yet, as the creative instinct is strong in the Celt, and it is the Celt who leads in art, there is no reason why in future years this strange Renaissance should not become almost as mighty in its way as was that new birth of Art that woke many centuries ago in the cities of Italy" (CW, 1050).

Throughout the eighties Wilde had moved freely in radical, socialist, and anarchist circles and had participated in or otherwise supported their causes. He knew intellectually, and sometimes personally, people who applied evolutionary theory to social problems: Spencer, Ruskin, Huxley, Clifford, E. B. Tylor, G. J. Romanes, J. S. Mill, Edward Carpenter, Sir Charles Dilke, George Eliot, James F. Stephen, Bernard Shaw, Prince Kropotkin, and Max Nordau. He was well aware of controversies concerning evolutionary theories of human improvement. He reviewed D. G. Ritchie's *Darwinism and Politics* (1889), a work which outlines and attempts to reconcile the political conflicts developed from materialist and idealist positions and theories of human development.[156]

From 1886 to 1894 *The Fortnightly Review*, edited by Frank Harris, provided a forum for the expression of radical Darwinist ideas. Like other radical Darwinists such as A. R. Wallace, Allen, Kropotkin, Ritchie, and Patrick Geddes, Wilde believed that science gave a picture of human nature and its development very different from the dominant Victorian view of a basically selfish, competitive, and brutal creature who improved through a bloody intraspecies struggle for existence caused by an inevitable scarcity of human necessities.[157] Radical Darwinists believed that individuals were naturally social and creative creatures who cooperated with their fellows to ensure the survival of the human species. In their opinion, the wasteful, inhumane, and brutally competitive "law of the jungle" which supposedly dominated human nature was encouraged by an artificially imposed social and economic system.[158]

In "The Critic as Artist" Wilde imagined a utopian "true culture" where aesthetics were higher than ethics. His analogy of aesthetics and sexual selection to ethics and natural selection showed his affinities with these radical Darwinists. In "The Soul of Man Under Socialism" he formulated a nonauthoritarian socialist society which would discourage economic competition and encourage contemplation and aesthetic activity (analogous to sexual selection) as the means of achieving continuous cultural and social improvement. By examining his review-essay "A Chinese Sage" (1890) and "The Soul of Man Under Socialism" (1891) in their intellectual and political contexts, we will suggest the coherence of Wilde's political and cultural theories as they developed directly from the synthesis of Hegelian idealism and evolutionary theory sketched in the notebooks.

We believe that critics have erred in interpreting "A Chinese Sage" and "The Soul of Man Under Socialism" as advocating social inaction.[159] It is true that Chuang Tsu's philosophy is quietist, but in Wilde's essay the Chinese philosopher became a Victorian sage with a dialectical and Darwinian philosophy. Wilde identified him with the philosophers and scientists whom he most admired, including Plato, Hegel, Rousseau, and Herbert Spencer. As for the "creed of inaction" and the virtues of contemplation which these critics see as the central doctrine of "The Soul of Man Under Socialism," we argue that Wilde's description and *use* of this creed reveal his awareness of its paradoxical social implications (which were apparent in his notebooks and earlier essays). After placing the Chinese philosopher in this contemplative and philosophical tradition, Wilde asked, "Who, then, according to Chuang Tsu, is the perfect man? And what is his manner of life? The perfect man does nothing beyond gazing at the universe. He adopts no absolute position. . . . He knows that, 'just as the best language is that which is never spoken, so the best action is that which is never done.' He is passive, and accepts the laws of life. He rests in inactivity, and sees the world become virtuous of itself."[160] Since the perfect man is not an absolutist, Wilde noted correctly that there was only "something in [Chuang Tsu] of the Quietist" (CS, 222). Wilde saw the paradox and so the irony of inaction and quietism when he remarked, "Chuang Tsu spent his life preaching the great creed of Inaction, and in pointing out the uselessness of all useful things" (CS, 222).[161] Chuang Tsu was quietist in his utopian social philosophy, not his personal activity: He had much to say about social issues, about government, economics, education, and philanthropy.

Wilde used Chuang Tsu to criticize the present "artificial society based upon capital" for its waste and destruction (CS, 224). In particular, Wilde criticized English misrule in Ireland, the priggish moralizing and altruism of philanthropists, and organized education (CS, 224–25). Instead of this activism, Wilde preferred an evolutionary quietism which would allow the world to become virtuous of itself. For, as Wilde pointed out, "Chuang Tsu is one of the Darwinians before Darwin. He traces man from the germ, and sees his unity with nature" (CS, 226). For Wilde and Chuang Tsu man in the state of nature embodied "original simplicity," and, if untampered with, "Virtue establish[es] itself" (CS, 227–28).

Therefore Wilde believed that "all modes of government are wrong. They are unscientific, because they seek to alter the original environment of man" (CS, 223).

"A Chinese Sage" is the first essay in which Wilde used evolutionary theory to support his radical critique of society. Far from describing a static utopia for contemplatives, he recommends inaction only in the context of a totally restructured society in which collective evolutionary improvement will occur because the naturally creative individual personality will be freed to perfect itself. His expanded proposals in "The Soul of Man Under Socialism" are well known, but scholars are less certain about the intellectual coherence of Wilde's argument. Is it contradictory to recommend both individualism and socialism? Can individuals really exist in groups without authority? Some scholars have, in the face of these seeming contradictions, expressed perplexity, while others have claimed that Wilde was ironic, or paradoxical, or both.[162]

We believe Wilde's paradoxes were intellectually witty but dialectically serious, based on the synthesis he developed in his notebooks and revised and incorporated in his cultural criticism. It had, we argue, four implications. The first was the establishment of a nonauthoritarian socialist polity and economy. "The Soul of Man Under Socialism" proposed a "scheme for the reconstruction of society" which, Wilde said, "trusts to Socialism and to Science as its methods" (CW, 1103). Second, his ideal society was individualistic and pluralistic: "There is no one type for man. There are as many perfections as there are imperfect men" (CW, 1087). Third, given this utopia, Wilde envisioned an ideal situation for autodidactic education for all persons. Fourth, he suggested that sexual selection (eugenics) would contribute to the continuing improvement of culture.

Wilde argued for both socialism and individualism, which were regarded in his own day as antithetical, because he saw them as interdependent elements necessary for human evolutionary progress. He began "The Soul of Man Under Socialism" by amplifying the critique of English society offered in "A Chinese Sage." The institution of private property destroyed the possibility of individual fulfillment for most of the rich and all of the poor. It produced a static and pathological system in which only a privileged few (such as Flaubert, Keats, Renan, and Darwin) could realize true individuality (CW, 1080). Wilde's goal was the realization for all people of this sort of individualism, and for this goal he claimed scientific authority: "To ask whether Individualism is practical is like asking whether Evolution is practical. Evolution is the law of life, and there is no evolution except towards individualism. Where this tendency is not expressed, it is a case of artificially arrested growth, or of disease, or of death" (CW, 1100–1101).

Such individualism could occur only after a revolution which would end an artificially created struggle for existence in society. Wilde did not describe how this revolution might occur, although he praised revolutionary violence and the rebellious and disobedient acts of the poor (CW, 1081, 1094). As an anarchist-socialist, Wilde probably believed that true revolutions began spontaneously, but he recognized the necessity for agitators and continued to oppose philanthropy as detrimental to rich and poor alike (CW 1079, 1082).[163]

Only after this revolution would Wilde's nonauthoritarian socialist state make rebellious and revolutionary personalities unnecessary, for at that time the entire social structure would be organized to release the individual's latent creativity: "Socialism, Communism, or whatever one chooses to call it, by converting private property into public wealth, and substituting cooperation for competition, will restore society to its proper condition of a thoroughly healthy organism, and ensure the material well-being of each member of the community. It will, in fact, give Life its proper basis and its proper environment" (CW, 1080). Since his utopian socialist society is a healthy organism, it cannot be static. In fact, since it is man's nature to create and develop, Wilde said, when humanity lands at utopia "it looks out, and, seeing a better country, sets sail. Progress is the realisation of Utopias" (CW, 1089). And to ensure "the full development of Life to its highest mode of perfection, something more [than socialism] is needed. What is needed is Individualism" (CW, 1080).

Socialism would be valuable only insofar as it ended competition for the means of subsistence and freed the individual to pursue and create the beautiful and the good (CW, 1088). Wilde believed, like other radical Darwinists, that humane qualities did not evolve through natural selection.[164] But by allowing all individuals to take their physical subsistence for granted, that is, to live aesthetically, Wilde wished to free those humane qualities which, like sexual selection, would give new forms, variety, and progress to life and so allow mankind to evolve naturally toward humane individualism. For Wilde the free individual was the agent of improvement, and his utopia encouraged creativity rather than quietism (creativity, of course, would be impossible without contemplation). His ultimate vision was pluralistic: "it does not matter what [man] is, as long as he realises the perfection of the soul that is within him" (CW, 1087). One may be a poet, scientist, student, philosopher, shepherd, dramatist, fisherman, composer, or priest, so long as he is "perfectly and absolutely himself" (CW, 1087).

Wilde's model for the perfect personality is the artist as Shelley defined it in "The Defense of Poetry," an imaginatively creative individual who "expresses everything" in work which he can enjoy because he creates with no concern for what other people want (CW, 1093).[165] "The note of the perfect personality," Wilde said, "is not rebellion, but peace. It will be a marvellous thing—the true personality of man—when we see it. It will grow naturally and simply, flowerlike, or as a tree grows. It will not be at discord. It will never argue or dispute. It will not prove things. It will know everything. And yet it will not busy itself about knowledge. It will have wisdom" (CW, 1084).

Two emphases emerge from this description. One is Wilde's use of biological metaphor, which is consistent through the essay: Human beings are plants. In the late nineteenth century, when eugenics was in flower, this metaphor was taken quite literally by many writers. Second, and more important, is Wilde's belief that knowledge, wisdom, and imagination are innate. Because he assumed the biological origins of culture, Wilde asserted that the perfection of personality required a freedom to develop according to "its own laws" (CW, 1085). That is, Wilde literally believed that "what a man really has, is what is in him" (CW, p. 1084), and that there was a "great actual Individualism latent and potential in mankind generally" (CW, 1083). Although "The Soul of Man Under Socialism" contains no references to it, racial consciousness is assumed to be the biological source of both imagination and wisdom. As Gilbert asserts in "The Critic as Artist," "[T]he imagination is the result of heredity. It is simply concentrated race-experience." "Culture," he adds, is made possible by the "transmission of racial experiences" (CW, 1041). Cultural perfection can therefore be achieved only by self-perfection, that is, in the absence of any external authority.

It should now be clear why Wilde stressed the interdependence of socialism and individualism. If a competitive social structure repressed the development of aesthetic, intellectual, and moral faculties which he regarded as innate, concentrated race-experience, then only under socialism could this "soul of man" emerge. The inheritance of culture would make cultural authority unnecessary.

Wilde's educational theory, as expressed in the notebooks, the lectures in America, "The Critic as Artist," and "The Soul of Man Under Socialism," is consistent with these premises. Wilde believed that formal schooling played only a small role in what should be a spontaneous yet guided awakening of the individual's inherent aesthetic sensibility. He therefore differed radically with Arnold's proposal for a centralized state education system which would teach the individual to choose the good. According to Wilde, "education is an admirable thing, but it is well to remember from time to time that nothing that is worth knowing can be taught" (CW, 1016). Wilde maintained that Plato's dialectical method provided an anti-authoritarian model for the growth of the critical intellect. He reiterated in this essay what he said in "The Critic as Artist," namely, that true education was a process of self-realization which began with personal pleasure and developed through a sense of one's own creative potential: "Pleasure is Nature's test, her sign of approval. When man is happy, he is in harmony with himself and his environment" (CW, 1104). More specifically, truly desirable and humane education would lead to self-realization through cultivation of an aesthetic sensibility and a critical intellect. Wilde agreed with Arnold that the present educational system emphasized the memorization of unconnected facts. Moreover, the essential congruity of Platonic forms with Wilde's idea of the innate aesthetic sense of the individual should be clear. Both Plato and Wilde proposed "that real love of beauty . . . is the true aim of education" (CW, 1049; CB, 40, 185). Perfection could be accomplished only autodidactically, "for the development of the race depends on the development of the individual, and where self-culture has ceased to be the ideal, the intellectual standard is instantly lowered, and often, ultimately lost" (CW, 1043).

When Wilde talked about the improvement of the race, he was willing (unlike Arnold) to carry his scientific theory to a logical (and biological) conclusion: He suggested eugenics, artificial sexual selection of the best racial traits, as another method of cultural improvement. In Wilde's socialist society voluntary choice of mates based on mutual attraction would replace the usual Victorian practice in which needs of survival, kinship and class alliance, or enhancement of property took precedence. Wilde never made a concrete proposal for a nonauthoritarian form of sexual selection, but his interest in it is representative of the ideas of an important and forgotten group of cultural critics. In the late eighties radicals who were aware of Weissmann's theory of the noninheritance of acquired characteristics came to doubt the efficacy of environmental change as a means of cultural improvement and instead began to place their faith in eugenics. At that time, according to A. R. Wallace, a "flood of articles in our best periodicals" appeared on the subject.[166]

Wilde probably would have sympathized with Grant Allen's and A. R. Wallace's eugenic plans.[167] Allen argued in his essays (one of which appeared in *The Fortnightly Review*) for free love as part of a eugenic proposal which encouraged women to choose for child-bearing purposes temporary mates from among the finest, healthiest, and most intelligent men. In an essay which appeared in *The Fortnightly Review* four months before "The Soul of Man Under Socialism," Wallace also outlined a nonauthoritarian socialist scheme for human improvement through sexual selection. Stating that education could not lead to permanent cultural improvement, Wallace suggested that, once removed from economic competition and given the freedom to choose mates, women would be attracted to men who embodied what Victorians called "the higher qualities," and the cumulative hereditary impact of that sexual selection would therefore improve the culture of the race (HS, 335, 337). In "The Critic as Artist" Wilde stated that both imagination and the critical temperament were inherited racial qualities, adding, "Under what conditions, and by what means, this temperament is engendered in race or individual, we will not discuss at present. It is sufficient to note that it exists" (CW, 1049). In "The Soul of Man Under Socialism" he argued that marriage and family should be abolished in favor of a freer and more beautiful love relationship between man and woman (CW, 1086). As part of his proposal for a socialist utopia, this suggestion is his contribution to the debate among socialists and cultural critics over the eugenic role of sexual selection in cultural improvement.

We believe, then, that Wilde's cultural theory does represent a new and significant turn in the history of cultural criticism. Although Raymond Williams quotes Wilde's statement that culture is made possible by the "transmission of racial experience" (C & S, 170), he does not see how crucially this theory affects Wilde's social writings. For instance, it resolves into a dialectical unity the apparent contradiction between socialism and individualism. Thus Williams misperceives Wilde's individualism as the "pursuit of an isolated aesthetic pleasure" and therefore concludes that the aesthetics "rests essentially on a denial of society" (C & S 172).

Wilde's cultural criticism, understood as a whole, does not advocate political quietism or the pursuit of isolated aesthetic experiences. Neither is it intellectually unsystematic or paradoxical (although, obviously, Wilde uses paradoxes rhetorically). Rather, his proposals for cultural improvement, based on his synthesis of idealist philosophy and evolutionary and ethnological theories of his day, represent a coherent plan in which socialism, self-education, and biology would function to perfect both individual and society. Both these beliefs and the radical programs they engendered make Wilde's work representative of an important strain of cultural criticism which, quite understandably in the light of recent history, has been ignored or rejected, but for that very reason it should not be misunderstood or forgotten.

History and Faith: Interpreting the Sonnets

Critics generally view "The Portrait of Mr. W. H." as Wilde's interpretation of Shakespeare's sonnets and their Renaissance context offered up in fictional guise. We see it as a further variation of the synthesis developed in the notebooks. It is also Wilde's attempt to dramatize his interest in the psychology of creation and influence, to write an allegory of the development of a modern idealist historical method of interpretation, and to imagine a Hegelian interpretation of Shakespeare, his sonnets, and their historical importance as representative types of the English Renaissance Weltanshauung. Finally, it constitutes the kind of criticism Wilde describes in *Intentions*, one that deepens the wonder and mystery of Shakespeare's sonnets and inspires the creative and critical efforts of others.

The first version of "The Portrait of Mr. W. H.," like other essays and fictions by Wilde, was published in a periodical (*Blackwoods*, 1889) and then enlarged. Unlike "The Truth of Masks," "The Decay of Lying," "The Critic as Artist," and *The Picture of Dorian Gray*, however, the revision (completed in 1893) was not published in Wilde's lifetime, perhaps because of the controversy over homosexual implications aroused by its first version and intensified by the publication of *The Picture of Dorian Gray*. In both versions Wilde advances the thesis that W. H., to whom Shakespeare dedicated the sonnets, was not the Earl of Pembroke but rather a boy actor named Willie Hughes. He acknowledged his belief in the thesis when he chastised W. E. Henley for not seeing "the wonder and beauty of my discovery of the real Mr. W. H."[168] "The Portrait of Mr. W. H." is neither a simple fiction nor an essay in Shakespearean scholarship. It is that especially Wildean mixed mode, "art-criticism."

In "The Critic as Artist" Wilde, through Gilbert, defined the highest form of criticism as a synthesis of the creative (lying) and the historically (i.e., critically) interpretive: "[T]he true critic can, if it be his pleasure so to direct or waste his faculty of contemplation, produce work that will be flawless in beauty and instinct with intellectual subtlety" (CW, 1027). To enact this theory in practice, Wilde reworked his earliest essay and adopted the dialogue and other distancing devices to fictionalize his later critical and discursive prose. He was caught between contradictory impulses: his wish to present a theoretically consistent cultural criticism and his desire to create a mysterious and beautiful aesthetic criticism which "reveals everything, because it expresses nothing" (CW, 1030). Another resolution of this contradiction was to write aphorisms, Delphic utterances that are both verbally amusing and aesthetically satisfying. Wilde practiced this art of the miniature best in *Intentions*, in some of his fairy tales, in the "Preface" to *The Picture of Dorian Gray*, in "A Few Maxims for the Instruction of the Over-Educated," in "Phrases and Philosophies for the Use of the Young," in "Poems in Prose," and in many of the one-liners of his plays.

In "The Portrait of Mr. W. H." Wilde artistically realized this theory of "art-criticism" (stated in *Intentions*) by embodying criticism in fiction to enact the creative-interpretive process itself. The story, which allegorizes the evolution of the critical spirit in the modern world, is a fictional parallel to Wilde's description of the development of the historical method in the "Rise." "Mr. W. H." also incorporates Wilde's Hegelian interpretation of Western art history as a triple flowering (the Greeks, the Renaissance, and the English Renaissance), which he developed in "The English Renaissance in Art" and "The Truth of Masks." Finally, it includes Wilde's newer interests in the psychological effects of creativity and art, which appeared in "The Decay of Lying" and "The Critic as Artist."

This story of the realization of the critical spirit and the historical method begins *in medias res* when the young narrator dines with a middle-aged gentleman friend, Erskine. The narrator's aesthetic justification of literary forgeries prompts Erskine's tale about Cyril Graham, his dead friend, who had a portrait forged to prove his strange theory about Shakespeare's sonnets. Erskine's story quite unexpectedly inspires the narrator's belief. He, in turn, studies the sonnets and revises Graham's interpretation, expanding his original insight by fleshing out the sonnets' relations to the lives of the individuals involved, to the history of the drama, and to the arts of the Renaissance and ancient Greece. The narrator sends the manuscript to Erskine, hoping that he will publish it and do justice to Graham's theory, but in doing so the narrator loses his belief in it; meanwhile, Erskine is reconverted by the letter. The narrator's new skepticism drives Erskine to devote his life to finding the evidence to prove the theory. Two years later Erskine sends a letter that he hopes will re-convert the narrator. In it he says that he could not verify the theory, but that he will commit suicide to prove his belief in it. At the end, the narrator learns that Erskine died of consumption, not suicide; then the narrator meditates on the theory's problematic truth and the strange fate of those who believed it.

We divide our consideration of this "art-criticism" into three parts. In the first we discuss the interpretation of the sonnets themselves. In the second we examine the narrative description Wilde uses to dramatize the psychology of interpretation and creativity. Finally, in the third we describe the development of the Hegelian method and theory of historical criticism.

The Interpretation of the Sonnets

Wilde's interpretation of the sonnets appears in three incremental, complementary sections. It originates with Cyril Graham's inspired biographical discussion of evidence in the sonnets. Cyril claims that Shakespeare wrote these passionate love poems to a young man "who was a really vital factor in the development of his dramatic art" (CW, 1153) and was also a commoner. Dismissing rival biographical theories and the notion that the poems were allegorical, he poses the question, "Who was he whose physical beauty was such that it became the very cornerstone of Shakespeare's art; the very source of Shakespeare's inspiration; the very incarnation of Shakespeare's dreams?" His answer is the "boy-actor of Shakespeare's plays" who played the female romantic and tragic roles (CW, 1156).

Cyril suggests that the love and friendship themes in the sonnets should be interpreted in terms of an artist's love and concern for his medium: dramatic art. The narrator explores this idea of a biographical germ in the sonnets and then develops a historical analysis of Renaissance drama: "I will take up the theory where Cyril Graham left it, and I will prove to the world that he was right'' (CW, 1161). He begins, as Cyril did, exploring the love and marriage themes in the sonnets themselves. In the narrator's revision of Cyril's interpretation, Shakespeare's love for W. H. is not so much an inspiration as the love of an artist for his artistic medium, like the musician's love for his instrument or the sculptor's love for ''some rare and exquisite material'' (CW, 1173). As an actor W. H. constituted the material medium for the drama (CW, 1174).

The narrator also explains and justifies the theme of male friendship in the sonnets as a ''vital factor in the new culture'' (CW, 1174) of the Renaissance by relating it to its classical roots: ''It is only when we realise the influence of neo-Platonism on the Renaissance that we can understand the true meaning of the amatory phrases and words with which friends were wont, at this time, to address each other'' (CW, 1175). He describes the friendships of Michelangelo and Tomasso Cavalieri, of Pico della Mirandola and Ficino, of Montaigne and Etienne de la Boétie as expressions of one soul's love for another, ''the worship of intellectual beauty'' (CW, 1175) which Ficino saw as ''the realisation of the Greek ideal,'' expressed in its fullest form by Plotinus (CW, 1177). Insofar as Willie Hughes was the ''visible incarnation of [Shakespeare's] idea of beauty,'' says the narrator, ''it is not too much to say that to this young actor . . . the Romantic Movement of English Literature is largely indebted'' (CW, 1177).[169]

In section three of ''W. H.'' the narrator finds historical support outside of philosophy and art history by exploring the material causes of Shakespeare's art. He begins by explaining the historical typicality of Willie Hughes as a boy-actor of female roles. Shakespeare, he says, developed the theme of ''the ambiguity of the sexes'' (CW, 1180) from the material conditions and limitations of his art: Since women were not allowed to act, Shakespeare used the male impersonator to develop his theme. Here Wilde suggests that artistic inspiration comes more from an art's material medium than from any idea. Furthermore, he believes that the ''essentially male culture of the English Renaissance found its fullest and most perfect expression by its own method, and in its own manner'' (CW, 1182). That is, the drama itself becomes the essential romantic and modern art form because it is the most material art form, the most concrete and sensuous, using ''all means at once'' to appeal to the eye and the ear (CW, 1173).

The narrator next addresses the meaning and importance of the ''dark lady.'' Again he examines historical evidence for her typicality: She, like other women of her class, fell in love with the theater and then fell, quite specifically, for Willie Hughes. For the narrator Shakespeare's relation to the lovers represents the typical struggle between a life of the spirit and the flesh. Shakespeare first feigns love for the lady to divert her attention from his friend, but, bewitched by his own words and acts, he becomes truly enthralled by her and by a shameful physical lust. Finally he rejects her for his soul's mate, the young actor. Again the narrator produces external evidence which ''curiously corroborated Cyril Graham's wonderful guess at the secret of Shakespeare's great romance'' (CW, 1189). By using both external evidence and a careful study of internal evidence (i.e., ''language, style and the like'' [CW, 1189]), the narrator dates the ''dark lady'' episode early in Shakespeare's life. His explanation, he believes, places ''the new interpretation of the Sonnets on something like a secure historic basis'' (CW, 1189). This critical exercise inspires him: ''Something made me feel certain that Willie Hughes had survived Shakespeare, and had fulfilled in some measure the high prophecies the poet had made about him, and one evening the true secret of his end flashed across me'' (CW, 1192). W. H. had gone to Germany carrying copies of Shakespeare's plays; he therefore became ''the first to have brought to Germany the seed of the new culture, and was in his way the precursor of the *Aufklärung* or Illumination of the eighteenth century'' (CW, 1192).

So the narrator's interpretation of the sonnets, which begins in Cyril's biographical speculation about the sources of Shakespeare's inspiration, is developed into a carefully researched historical analysis of their representative status as manifestations of the Renaissance zeitgeist and the place of their actors in the development of the Enlightenment and the Romantic movement. Both inspiration and historical research figure crucially in Wilde's Hegelian explanation of the Elizabethan Renaissance and the origin of the nineteenth-century English Renaissance.

The Psychology of Influence, Expression, and Interpretation in "The Portrait of Mr. W. H."

The production of art (or ''art-criticism'') as a way of realizing and expressing the personality is a familiar feature of Wilde's artistic and critical theory and his utopian social philosophy. In ''The Soul of Man Under Socialism'' he imagines an evolutionary ''New Hellenism'' of pleasure, individual self-perfection, and expression of personality for all, not just for artists and critics. In *De Profundis* Wilde sees individual self-expression through sorrow as a development of his previous position in ''The Soul of Man Under Socialism,'' but he holds to the idea (a repeated motto in the letter), that ''the supreme vice is shallowness. Whatever is realised is right'' (*Letters*, p. 448). In ''The Picture of Mr. W. H.'' Wilde designs his narrative to demonstrate how personal and cultural influences occur through the subjective shaping of the interpretive process: He describes the transformation of spirit from inspiration to self-conscious objectification. Again, Wilde's Hegelian-evolutionary synthesis, most evident in the narrator's analysis of his experiences of influence and expression, provides the intellectual framework for his tale of the realization and transmission of influence through personal contact (dialogue, even discipleship) and through mediation (a narrated story, a work of art or criticism). His discussion of soul and its relation to individual consciousness shows the importance of both Spencer's and Clifford's conception of an inherited racial unconscious.[170] Looking back on his Oxford years, the narrator now realizes that he owes his ''passion for Greek thought and art'' and his ''sympathy with the Hellenic spirit'' to the ''influence'' of a ''friend'' he had known for many years but never seen. That friend is, of course, his soul:

> There had been a presence beside me always. . . . What though I had been unconscious of it? The soul had a life of its own, and the brain its own sphere of action. There was something within us that knew nothing of sequence or extension, and yet, like the philosopher of the Ideal City, was the spectator of all time

> and of all existence. It had senses that quickened, passions that came to birth, spiritual ecstasies of contemplation, ardours of fiery-coloured love. It was we who were unreal, and our conscious life was the least important part of our development. The soul, the secret soul, was the only reality. (CW, 1195)

Here influence is a dialectical process in which an unrealized aspect of the collective soul is brought to awareness by an outside agent: an individual, or some form of artistic or cultural expression. In "The Picture of Mr. W. H.," as in all Wilde's mature criticism, the proper function of art is to provide the external spiritual agency that will allow the individual's unconscious to realize itself, to discover the soul comprised of racially inherited experience: "Art, even the art of fullest scope and widest vision, can never really show us the external world. All that it shows is our own soul, the one world of which we have any real cognizance" (CW, 1194).[171] Wilde's conception of the soul resembles Jung's idea of the collective unconscious: It is a storehouse of racial and cultural memories potentially alive in everyone and passed on through heredity, enriched or impoverished by each individual's experiences.[172] For Wilde the Greek ideal of self-knowledge meant the realization of that unconscious culture.

This spiritual awakening could occur through a dialectic of personal relationships. In the story the narrator says: "Influence is simply a transference of personality, a mode of giving away what is most precious to one's self, and its exercise produces a sense, and, it may be, a reality of loss. Every disciple takes away something from his master" (CW, 1196). Wilde uses the spiritual influences of art and individuals to organize the plot of the story: Shakespeare's relationship to Willie Hughes is the prime example of such influence. As person and actor, W. H. has inspired Shakespeare (and also Marlowe) to write poems and plays. These, in turn, are the spiritual and efficient causes of the rest of the story's action. They inspire Cyril's interpretation of the sonnets, which, with his martyrdom, inspire the narrator's revised and expanded interpretation. He rereads Shakespeare's work and is inspired by it to write a historical, idealist interpretation of the Renaissance that focuses on the spiritual dialectic of personal and artistic influences and their inspiration in ancient Hellenism. This interpretation reinspires Erskine to search further for empirical proof of the historical Willie Hughes.

Wilde's portrayal of influence in "The Portrait of Mr. W. H." also includes the psychology of transmission. He focuses on the process of self-expression when, in a passage that echoes an idea in "The Critic as Artist,"[173] the narrator explains his catharsis after sending Erskine the interpretation:

> No sooner, in fact, had I sent it off than a curious reaction came over me. It seemed to me that I had given away my capacity for belief in the Willie Hughes theory of the Sonnets, that something had gone out of me, as it were, and that I was perfectly indifferent to the whole subject. What was it that had happened? It is difficult to say. Perhaps, by finding perfect expression for a passion, I had exhausted the passion itself. Emotional forces, like the forces of physical life, have their positive limitations. Perhaps the mere effort to convert any one to a theory involves some form of renunciation of the power of credence. (CW, 1196)

The narrator's loss, however, becomes the recipient's inspiration, a transference of spirit.

In his portrayal of influence Wilde also dramatizes a dialectical psychology of interpretation. Cyril's interpretation and the narrator's development of it, while having a common core, are also expressions of their own subjectivity.[174] Cyril is a modern Willie Hughes, a pleasure-loving, beautiful, effeminate young man who, while at Cambridge, performed female roles in Shakespeare's plays. Cyril's interpretation is an inspired, naive tour de force which, not surprisingly, claims that the sonnets focus on, and were inspired by, Shakespeare's love for a young and beautiful man "who was a really vital factor in the development of his dramatic art" (CW, 1153). Relying, as Erskine says, "on a kind of spiritual and artistic sense" (CW, 1156), Cyril ignores large sections of the sonnets and reasons ingeniously to project his personality and desires into the poems.

The narrator does not have an Erskine to describe his character, but he distinguishes himself from the impulsive and enthusiastic Cyril through his opinions, his way of developing an interpretation, and his narrative style. Like Cyril, he is an affluent, well-educated young man interested in the arts. While Cyril was "wild to go on the stage" (CW, 1153) and did not hesitate to use forgery to prove his theory, the narrator presents himself as a critic with carefully reasoned opinions (not unlike Wilde's) about art; for instance, at the beginning of the story he discusses forgery in creative and historical work. His interpretation, while inspired, is developed fully only after months of copious reading in creative and scholarly sources and represents a broader, more historical perspective on the sonnets. The narrator builds upon and encompasses Cyril's theory, but he shifts the focus of the sonnets from Willie Hughes to the more mature Shakespeare. He finds "the true interpretation, which indeed Cyril Graham himself seemed to have missed" (CW, 1164), in the mystery of the marriage metaphor, which he interprets and sublimates as the artist's concern with the medium of his art. For the narrator W. H. is important because he inspired Shakespeare to write sonnets which were historically representative of the spirit of the Renaissance.

By tracing the evolution of this interpretation, Wilde explores individual psychology within the larger context of a dialectical and racial conception of cultural development. Cyril's and the narrator's interpretations are shaped by their differing personalities, but both represent aspects (intuitive and critical) of the new Hellenic temper, the union of imaginative and historical understanding that constitutes truth in Hegelian terms. Both are interested in and inspired by Renaissance art. Both are creative and capable of influencing others by their words. Both, in their differing ways, sustain a belief in the truth of their theories despite lapses caused by self-expression and a demand for proof positive of Willie Hughes.

Erskine's place in this dialectic is more problematic. Although he admires art and is inspired by Cyril and the narrator, his background and actions suggest a temperament antithetical to theirs. Educated in the sciences at Cambridge, the son of a Radical and probably Utilitarian father, he demands independent, external evidence to prove the truth of Cyril's theory. As a cultural type he suggests the kind of pragmatic, educated Briton who would be fascinated by his flamboyant schoolboy friend Cyril and yet would be reproached by him for his "philistine tone of mind" (CW, 1158). On the issue of faith, which he cannot concede to the narrator's theory, he shows his difference with the two. He is not only bound to empiricism, he is a copier. The conclusion of the story reveals his fraud (copying

Cyril), intended to reconvert the narrator to belief in Willie Hughes. However, Erskine can only accept and reproduce the theory, not create a new version; the fraudulent suicide is simply copying Cyril, and the revelation of his fraud (the fact that he died of consumption) shows the inauthenticity of his effort to influence the narrator. Like the materialists in the notebooks (CN, 57–63) and the realists in "The Decay of Lying," Erskine can neither realize the importance of soul nor find in his own soul the wisdom which for the narrator constitutes a modern faith.

THE MODEL OF A MODERN HISTORICAL METHODOLOGY IN "MR. W. H."

> Further, as regards the particular method of investigating that group of phenomena obtained for him by the abstract method, he [Polybius] will adopt, he tells us, neither the purely deductive nor the purely inductive mode but the union of both. In other words, he formally adopts that method of analysis upon the importance of which I have dwelt before.
>
> "The Rise of Historical Criticism" (CW, 1138)

The narrator's wisdom at the end of the story arises partly because he recognizes the need to synthesize reason, inspiration, and evidence to understand the past. The third interest of "The Portrait of Mr. W. H." is Wilde's allegory of the development of this synthetic method, which he first described in the "Rise." There Wilde integrated his idealist belief in the spirit with the assumptions of science by suggesting that the basis for any scientific procedure, the "law of uniform causation," originated as an inspiration to the Greeks. Good historical practice depended crucially on the historian's imaginative ability to select and interpret facts, which were manifestations of the spirit existing in harmony with empirical research.

The problematic aspect of this method lies in the right relation between inspiration and fact: Wilde makes this relationship (and the development of a method which embodies it) the intellectual focus of his story. Cyril's intuitive interpretation represents a modern version of the spontaneous wisdom Wilde attributes to the ancient Greeks. Erskine's skepticism and demand for evidence corresponds to the naive materialist thinkers who questioned early Hellenic religious myths. As the story proceeds, these two forces become the focus of Cyril's, Erskine's, and the narrator's disagreements over interpretation. Only at the end of the story does the narrator realize in theory and practice a modern historical method (analogous to the Greek historians') which synthesizes idealism and empiricism.

The story introduces this problem when the narrator meets Erskine, who asks, "What would you say about a young man who had a strange theory about a certain work of art, believed in his theory, and committed a forgery in order to prove it?" (CW, 1150). The narrator had previously defended forgery (or lying) in art, but now he agrees with Erskine that when used for historical proof it is "quite a different matter" (CW, 1150). Erskine recounts how Cyril believed he had discovered the sonnets' "true secret": "This was Cyril Graham's theory, evolved as you see purely from the Sonnets themselves, and depending for its acceptance not so much on demonstrable proof or formal evidence, but on a kind of spiritual and artistic sense, by which alone he claimed could the true meaning of the poems be discerned'' (CW, 1156). His interpretation, that is, has both intuitive origin and validation, and in developing it he works ''purely by internal evidence'' (CW, 1153).

Cyril thus recapitulates the ancient Greeks' spontaneous realization of the idea of uniform sequence: ''[I]t seems to have come to the Greek mind spontaneously, through a certain splendour of imagination . . . before inductive research had armed them with the instruments of verification'' (CW, 1122). However, like Cyril, the Greeks applied the idea only haphazardly until Polybius, the ''culmination of the rational progression of Dialectic'' (CW 1133), developed it into a methodology. Just as skeptics challenged early Greek idealists, so the empiricist Erskine challenges Cyril; his inability to provide the kind of evidence Erskine desires ends the first dialectical encounter.

Cyril cannot imagine the need for historical research to convert Erskine. When he sees the forgery as external evidence, he shows both the power of his belief and his inability to do that research. Cyril's forgery, his ''lying'' confuses the aesthetic and scientific realms, just as Erskine's rejection of the theory because of the forgery confuses ethical with historical judgment. But Cyril' self-martyrdom, a proof of his belief, has the appropriate inspirational effect on the narrator. ''Cyril Graham,'' he says, ''sacrificed his life to a great idea . . . and if you will not tell of his martyrdom, tell at least of his faith'' (CW, 1161). The narrator steps into Cyril's role, hoping to convert the skeptical Erskine and the world. Erskine notices the resemblance when he remarks. ''Ah! how you remind me of Cyril!'' (CW, 1161).

In the second section of the story the narrator accepts the core of Cyril's theory and, relying heavily on external evidence, writes his Hegelian, historical interpretation of the sonnets and the Renaissance. That interpretation, which synthesizes intellectual and material sources, Neoplatonic philosophy, and the medium of Shakespeare's art (stage conventions and actors), represents the relationship of idea and fact in historical method that Wilde first defined in his notebooks.[175] In recognizing that the dramatist's material medium gives life and substance (as well as inspiration) to his fancy, the narrator reflects his own subjective desire and self-imposed task, namely, to give historical substance and reality to Cyril's inspiration. Confronted by Erskinian skepticism, the narrator turns to history to find support for his intuitive knowledge. He provides scholarly reasoning and extrinsic evidence which show the historical typicality of the sonnets' themes and of Shakespeare's relationship with W. H. He explains the ''dark lady'' sonnets by referring to documents which clarify the relations of Shakespeare, W. H., and the woman. His project, rooted in a subjective response to Cyril's theory, becomes at once an objective development of an idealist, dialectical, historical method and an intuitive and unselfconscious enactment of it.

As in the ''Rise,'' true historical understanding comes only with a fully self-conscious methodology. Wilde, preparing for this revelation at the story's conclusion, withholds self-awareness from the narrator, keeping him unconcerned with the gap between Hegelian and empirical tests of validity. In an ironic confrontation Erskine completes the narrator's interpretation and self-consciousness by adding one historical fact and a point about the logic of science. The narrator (now the lapsed believer) insists that the entire theory fails without empirical proof for the historical Willie Hughes. Erskine provides support for Cyril's original contention that the absence of W. H.'s name from the First Folio actors' list is a kind of proof of his existence: Given the public prejudice against actors and his desertion of Shakespeare, it was "extremely probable" that Willie Hughes would have adopted a stage name. Since such behavior was historically typical for actors, Erskine concludes, "it seems to me the most simple and rational explanation of the matter, I regard Cyril Graham's theory as absolutely proved" (CW, 1198). Erskine has proved the historical case by providing what might be called Hegelian, rather than empirical, evidence: Absence implies presence because it was probable and therefore typical that W. H. would adopt a nom de plume. When challenged by the narrator to provide evidence, Erskine gives the theoretical rationale for its scientific validity: "I feel quite sure that my theory is the true one. Of course it is a hypothesis, but then it is a hypothesis that explains everything, and if you had been sent to Cambridge to study science, instead of to Oxford to dawdle over literature, you would know that a hypothesis that explains everything is a certainty" (CW, 1199). The narrator realizes that scientific theory and Hegelian philosophy are one because neither requires total confirmation through empirical evidence and both require a kind of faith which accepts probability as sufficient proof.

In the final development of the allegory the narrator ponders the significance of Erskine's lie about suicide. This act, which conflates Cyril's suicide and his forgery and is committed for the same reason, recalls the historical/aesthetic problem Erskine poses for the narrator at the beginning. Erskine's effort to reconvert the narrator by "the pathetic fallacy of martyrdom" fails: "I had grown wiser since I had seen him. Martyrdom was to me merely a tragic form of scepticism, an attempt to realise by fire what one had failed to do by faith" (CW, 1201).

Wisdom, as Wilde wrote in the Commonplace Book, is the end of contemplation (CB, 189). Wisdom leads the narrator to see Erskine's act as a double instance of skepticism, a lack of faith. In holding this position, he also redefines faith: "No man dies for what he knows to be true. Men die for what they want to be true, for what some terror in their hearts tells them is not true" (CW, 1201). Faith, for the narrator, sustains life, and his ultimate position on the W. H. theory shows that he has evolved a faith appropriate to the modern age, one, that is, which can be reconciled with the truths of historical fact.

The narrator, relying on dialectical wisdom, avoids Cyril's and Erskine's mistakes by synthesizing faith and facts. He rejects Cyril's portrait as historical proof of the W. H. theory, but he retains it as a work of art. He also avoids Erskine's error of rejecting a historical theory on moral grounds. He unites Cyril's inspired sense of truth with Erskine's rigorous demand for external evidence and his theoretical understanding of the indeterminacy of scientific and historical truth. The story's last lines capture the balance of inspiration and evidence as well as the quality of belief appropriate in matters of historical interpretation. Sometimes, says the narrator, "when I look at it [the portrait], I think there is really a great deal to be said for the Willie Hughes theory of Shakespeare's Sonnets" (CW,

1201). The portrait, now art rather than evidence, inspires the narrator, who, lacking Erskine's and Cyril's fervency, recognizes that their faith and self-sacrifice are outworn and misplaced, and that his modern faith must be reconciled with the inevitable uncertainty that reason and the historical method demand. Wilde's story is, of course, "art-criticism," a form which problematizes his interpretation. The story, like the portrait, is an artifact admirable in its own right, but one that may also tell the truth about the origin and meaning of Shakespeare's sonnets.

Conscience as Tribal Self in *The Picture of Dorian Gray*

The Picture of Dorian Gray also appeared in more than one version, a fact that offers both opportunities and perils for critics. The book was first serialized by *Lippincott's Magazine* (1890); later, under pressure to lengthen the manuscript, Wilde added material that filled out ideas and characterizations, and at times apparently contradicted the earlier text. He also added the aphorisms that make up "The Preface." We will not explain away all the textual and interpretive puzzles, but we believe that Wilde's synthesis and the problems it addresses provide an intellectual ground which will clarify some of the book's apparent inconsistencies, its structure, and Wilde's intentions. The opposition of idealism and materialism in philosophy, of altruism and egoism in ethics, the relations of mind and brain, of soul and body, the role of lying in life and art—these issues, along with dialectical, evolutionary, and racialist conceptions of perception, of the psychology of artistic and nonartistic influence, flesh out the skeleton of a melodramatic plot.

While critics have mined the novel for unconscious, unintended meanings and for tantalizing autobiographical clues about homosexuality, Wilde consciously intended a meaning that he stated more than once, and quite fully in this letter:

> [I]t is a story with a moral. And the moral is this: All excess, as well as all renunciation, brings its own punishment. The painter, Basil Hallward, worshiping physical beauty far too much, as most painters do, dies by the hand of one in whose soul he has created a monstrous and absurd vanity. Dorian Gray, having led a life of mere sensation and pleasure, tries to kill conscience, and at that moment kills himself. Lord Henry Wotton seeks to be merely the spectator of life. He finds that those who reject the battle are more deeply wounded than those who take part in it. Yes; there is a terrible moral in *Dorian Gray*—a moral which the prurient will not be able to find in it, but which will be revealed to all whose minds are healthy. Is this an artistic error? I fear it is. It is the only error in the book. (*Letters*, 259)

If Wilde was uneasy about the aesthetic place of didacticism in fiction, he was not at all apologetic about the ideas themselves. While some readers have seen in the novel's moral and conclusion a Puritan, antiaesthetic strain, there is ample evidence from the notebooks and later work to confirm the fact that Wilde's conscious ethical norm was Aristotelian: the harmonious development of all one's powers. To love beauty too much, to be a detached and passive observer of life, and to pursue a life of "mere sensation and pleasure" are errors in a story about the destructiveness of obsessions or one-sided development.

In another letter Wilde opposes the consequentialist ethical tradition shared by nineteenth-century Puritans and utilitarians by stating a corollary of his Aristotelian, idealist ethics: "Right and wrong are not qualities of actions, they are mental attitudes relative to the incompleteness of the ordinary social organism. When one contemplates, all things are good" (*Letters*, 265; see CB, 185). For Wilde the ethical is defined as a state of mind. The contemplative attitude (*theoria*), the keystone of Aristotle's *Ethics* as Wilde understood it, is the defining characteristic of the good, the means and the end of self-perfection, of realizing one's *arete* (CB, 189). Wilde's criticism of Lord Henry's flawed version of contemplation—"to be merely the spectator of life"—shows again his desire to distance himself from Pater's more passive and hedonistic position in "The Conclusion" to *The Renaissance*.

In the same letter Wilde defines his ideal mental state: He looks forward to the time when a "sense of beauty will be the dominant law of life" (*Letters*, 265). The sense of beauty, the ability to perceive and love it, is the ideal good. The three major characters in *Dorian Gray* have lost, or lose during the course of the novel, this ideal mental state, the sense of harmony and of harmonious development that Wilde attributed to the ancient Greek mind as the basis for scientific and artistic achievement.[176]

The "terrible moral" to *Dorian Gray* grows out of Wilde's belief in a spiritual reality manifest in conscience and art. Dorian learns it as a result of his experiment in living and states it to the skeptical Lord Henry near the end of the novel: "The soul is a terrible reality" (CW, 161). The issue of faith versus doubt separated Victorians into two intellectual camps in philosophy, social ethics, and psychology. As a believer and philosophical idealist, Wilde accepted the reality and power of mind, acting imaginatively through language, to shape personal life and historical evolution and to act upon ideas. As a believing idealist, Wilde rejected the Paterian (and so utilitarian) hedonist's assumption of the individual ego's privileged position in psychology and ethics, as well as in social and economic matters; likewise, he rejected Pater's paradoxical denial of free will in favor of scientific determinism. We argue that two theories Wilde recorded in the notebooks best explain the action and meaning of the novel: the inheritance of acquired characteristics and Clifford's related belief in a tribal self, a powerful social consciousness which preceded the development of the ego. Indeed, Wilde structured the novel on the opposition between his idealist, Hegelian aesthetic humanism and Pater's materialist, aesthetic hedonism; its concluding scene resolves the question of the soul's existence and power.

Wilde saw materialist and idealist positions as manifestations of mind in dialectical relation (thesis and antithesis). The novel's conflict develops between Lord Henry's often cynical and reductive assertions of materialism and Basil's idealist protests and warnings against it. Basil and Lord Henry introduce their ideas in the context of a struggle for the attention and friendship of Dorian, who is (and whose name suggests) Wilde's ideal, "what I would like to be—in other ages, perhaps," an analogue to the innocent perfection of ancient Greece (*Letters*, 352). Dorian converts to, and acts on, Lord Henry's philosophy, but evolutionary idealism finally triumphs through the dialectical interaction of Basil's portrait with Dorian's racially transmitted imagination and conscience.

Basil has found the inspiration for his masterpiece, which he believes expresses the secret of his soul, in Dorian's youthful beauty, much as Shakespeare was inspired by W. H., much as Wilde in his critical writings suggests that new forms and motives can revitalize art. Dorian's influence has enabled Basil to integrate the soul and the body, the romantic passion and Hellenic perfection kept separate in earlier art. However, this influence is one-sided. Basil idolizes Dorian's beauty but inspires nothing more than polite friendship—Dorian is an unmoved mover.

When these three characters first meet, Basil introduces three aspects of Wilde's aesthetic Hellenism: the Hegelian description of art history and art's power, the identification of the artist's soul with the expressive power of his art, and the abstract sense of beauty (CW, 24–25). Lord Henry offers instead a materialism based on "the wild struggle for existence" (CW, 25), a "new Hedonism" (CW, 32), which he implies would be "finer, richer, than the Hellenic ideal" (CW, 29). The narrator paraphrases Pater to describe Lord Henry's doctrine: "Yes: there was to be, as Lord Henry had prophesied, a new Hedonism that was to recreate life. . . . It was to have its service of the intellect, certainly; yet, it was never to accept any theory or system that would involve the sacrifice of any mode of passionate experience. Its aim, indeed, was to be experience itself, and not the fruits of experience" (CW, 104). Wilde presents Lord Henry's materialism through his talk, actions, and inaction. As a thoroughly modern thinker, Lord Henry bases his "new Hedonism" on a strictly materialist premise: "Life is not governed by will or intention. Life is a question of nerves, and fibres, and slowly built-up cells in which thought hides itself and passion has its dreams" (CW, 162). He denies the existence of soul, claiming: "Good resolutions are useless attempts to interfere with scientific laws. Their origin is pure vanity. Their result is absolutely *nil*" (CW, 84). For him morality is a rationalization and crime is "simply a method of procuring extraordinary sensations" (CW, 160). The ethic he espouses is the individual's pursuit of pleasure: egoistic gratification. He also attributes the altruistic impulses of conscience to self-regarding motives: "Conscience and cowardice are really the same things, Basil. Conscience is the trade name of the firm" (CW, 22). Fidelity in love is "purely a question for physiology. It has nothing to do with our own will" (CW, 37). Nevertheless he tells Dorian that "from the scientific point of view" he should not try to influence Dorian's behavior since "all influence is immoral" (CW, 28).

Obviously Lord Henry sometimes cannot see the contradictions between his ideas and his life, but occasionally he is consciously deceptive, that is, a liar. Basil is only one character who points out the contradictions in Henry's life: "You are an extraordinary fellow. You never say a moral thing, and you never do a wrong thing" (CW, 20). Basil, as the novel shows us, is too kind. Lord Henry's hypocrisy works the other way, too. Although he claims that influence is immoral, he thinks to himself, "There was something terribly enthralling in the exercise of influence. . . . Yes, he would try to be to Dorian Gray what, without knowing it, the lad was to the painter who had fashioned the wonderful portrait. He would seek to dominate him . . ." (CW, 41). Here Lord Henry is a conscious hypocrite: His actions and decisions contradict his stated ethical and scientific ideas. At times the artist and the scientist must practice an amoral, total detachment; but Lord Henry also confuses the ethics of science and art with the ethics of life by playing Pygmalion to Dorian. He excuses lying to Dorian by thinking of him as one who "could be fashioned into a marvellous type," and later he muses to himself that "to a large extent the lad was his own creation" (CW, 41, 55).

Lord Henry rationalizes as Neoplatonic idealism his desire to become an artist in life through personal influence; but, compared to the narrator of "W. H." or to Basil, Lord Henry exploits his relationship with Dorian for dominance and self-interest, not for love. His materialism contradicts the liberating Hellenic spirit, "to philosophize with love"—that Greek phrase the narrator of "W. H." remembers stirring him in his Oxford days (CW, 1195).[177] Even when Wilde puts ideas from "The Soul of Man Under Socialism" or "The Critic as Artist" in Lord Henry's mouth, we cannot see Lord Henry as Wilde's spokesman. Lord Henry cynically appropriates anything that suits his purposes. He has the courage of neither his own—nor others'—convictions.

This relation of personal motive to idea embodies Wilde's belief that right and wrong are mental states rather than acts. Lord Henry preaches Hellenism, artistic influence, and even the practice of the scientific method: "[T]he ordinary subject-matter of that science had seemed to him trivial and of no import. And so he had begun by vivisecting himself, as he had ended by vivisecting others" (CW, 55). In fact, he identifies the method of modern science with the practice of contemplation in a passage which shows that he understands the ethical aspect of neither: "I don't desire to change anything in England except the weather . . . I am quite content with philosophic contemplation. But as the nineteenth century has gone bankrupt through an over-expenditure of sympathy, I would suggest that we should appeal to Science to put us straight. The advantage of the emotions is that they lead us astray, and the advantage of Science is that it is not emotional" (CW, 44). As Wilde's metaphor of vivisection suggests, Lord Henry desires to inflict pain and rationalizes his desire by confusing the ethics appropriate to the realms of science and life.

Lord Henry also will not publicly acknowledge the power of words and ideas. Basil, as an idealist, recognizes that power and warns Henry and Dorian that Henry's words have dangerous effects. The narrator, too, makes it clear that the power of words to express mind converts Dorian: "The few words that Basil's friend had said to him—words spoken by chance, no doubt, and with wilful paradox in them—had touched some secret chord that had never been touched before, but that he felt was now vibrating and throbbing to curious pulses. . . . Words! Mere words! How terrible they were! How clear, and vivid, and cruel! One could not escape from them. And yet what a subtle magic there was in them! They seemed to be able to give a plastic form to formless things. . . . Mere words! Was there anything so real as words?" (CW, 29–30).

Lord Henry has similarly cynical relations with people and ideas; he believes in neither and manipulates both to maximize his own satisfaction. As he says to

Dorian, "Faithfulness is to the emotional life what consistency is to the life of the intellect—simply a confession of failures" (CW, 50). In that sense he argues for—and Dorian lives out—Pater's egoistic ethic: He accepts no "theory or system that would involve the sacrifice of any mode of passionate experience" (CW, 104).[178] So, although he gave Lord Henry most of the witty lines, Wilde did not approve of him. In another letter he defined his relation to Lord Henry and Basil: "Basil Hallward is what I think I am: Lord Henry what the world thinks me" (*Letters*, 352).

We agree with Rodney Shewan that Wilde's remark about his relation to these two characters and also to Dorian ("What I would like to be—in other ages, perhaps" [*Letters*, 352]) is an important point of entry for critics seeking historical and biographical interpretations. "This succinct account," he writes, "forms the basis of my argument," which is that "the book itself is really a triple portrait of the artist as a young man" (Shewan, *Art and Egotism*, 112–13). Shewan gives Basil, Lord Henry, and Dorian equal places in this autobiographical triad and claims that "Wilde's enduring dilemma" of egoistic romanticism at odds with critical scepticism "undergoes definitive resolution" in the novel (116). We believe, however, that Shewan uncharacteristically and crucially misreads Wilde's words. When Wilde describes Lord Henry as "what the world thinks me," not "what I think I am," he is distancing himself from that character. Obviously Wilde created all of his characters, but we disagree with Shewan's suggestion that Wilde identifies himself with Lord Henry in the same way that he does with Basil. He gives Lord Henry the characteristically self-gratifying ideas and positions of a Paterian hedonist because that *was* what the world thought him. However, even as Lord Henry seemingly triumphs over Basil by converting Dorian, his ideas and acts are verbally or dramatically undercut from an idealist perspective on language, morality, and the soul.

The idealist response to Lord Henry comes first from Basil, whose thought and art represent Wilde's idealist synthesis. Basil's imaginative and contemplative power give him the dialectical awareness needed to create formally beautiful art and to see the spirit manifest in it: "The reason I will not exhibit this picture," he tells Lord Henry, "is that I am afraid that I have shown in it the secret of my own soul" (CW, 21). Basil enacts the Hegelian dialectic of creation and produces in the portrait a work of art which, as Hegel suggests, conveys spiritual truth in arresting sensuous form.[179] Another comment Basil makes about the potential artistry of Sybil Vane reveals much of his idealism: "To spiritualise one's age—that is something worth doing. If this girl can give a soul to those who have lived without one, if she can create this sense of beauty in people whose lives have been sordid and ugly, if she can strip them of their selfishness and lend them tears for sorrows that are not their own, she is worthy of all your adoration, worthy of the adoration of the world" (CW, 71). Such a theory, namely, that with the power of words one can give a soul to, or create a sense of beauty in, a people, is not an orthodox Christian belief but rather an optimistic expression of what the inheritance of acquired characteristics might accomplish. Basil's definition of the ideal soul—a "sense of beauty"—repeats Wilde's own ideal in the letter quoted earlier (*Letters*, 256).

Basil, the creator of the painting that becomes the embodiment of Dorian's conscience and soul (CW, 79), loves and loses in the early struggle for that soul. Basil's imaginative idealism is antithetical in every way to Lord Henry's ideas, and his actions embody the Neoplatonic ideal of spiritual love. Dorian recognizes it and gauges its worth in an episode of momentary remorse and insight: "Basil would have helped him to resist Lord Henry's influence, and the still more poisonous influences that came from his own temperament. The love that he bore him—for it was really love—had nothing in it that was not noble and intellectual. It was not that mere physical admiration that is born of the senses, and that dies when the senses tire. It was such love as Michael Angelo had known, and Montaigne, and Winckelmann, and Shakespeare himself" (CW, 97).

In fitting contrast to Lord Henry's lying and manipulation, Basil expresses his affection for Dorian only in and through the spiritual meaning of the painting. In an act that ironically foreshadows Dorian's last moment, Basil tries to destroy his own painting to prove to Dorian that their friendship means more than his masterpiece. But Dorian, who will later kill Basil and stab the portrait, intercedes to save it—not out of friendship for Basil but because he is in love with his own image.

Dorian's role in the story is to be an object lesson, to represent what hedonistic materialism might do to Western civilization. Before Dorian met Henry, he was to Basil an unselfconscious idealist and motive for art, "the visible incarnation of that unseen ideal whose memory haunts us artists like an exquisite dream" (CW, 93), that is, an embodiment of Hellenic racial memory.[180] Under Henry's tutelage he abandons his belief in the soul and lives by a materialist ethic. His character then develops a contradiction between his conscious and unconscious ideas, and his relation to his soul's heritage becomes thoroughly problematic.

After embracing materialism, Dorian is confronted with the reality of free will: At crucial moments he must choose between right and wrong. In each case he instinctively knows the right but is persuaded to do otherwise, as his response to Sybil Vane's death shows. With each evil act the soul he denies is transformed, made poisonous, and he becomes increasingly self-centered. Finally he realizes the inadequacies of Henry's theories and questions "the shallow psychology of those who conceive the Ego in man as a thing simple, permanent, reliable, and of one essence. To him, man was a being with myriad lives and myriad sensations, a complex multiform creature that bore within itself strange legacies of thought and passion . . ." (CW, 112).

Here Wilde forthrightly rejects ego psychology in favor of Spencer's and Clifford's view of the human psyche as a collective inheritance of acquired characteristics. In the following passage he shows Dorian pondering the unconscious influence of his ancestors and the "more absolutely conscious" influence of literature on the making of his soul (CW, 113). Dorian may understand the principle of exploring the inherited "strange legacies" of thought and passion, but he also chooses to explore only the *evil* aspects of family and cultural background. Thus, in his repeated ambles through the "gaunt cold picture-gallery of his country house," he finds himself fascinated by the ancestral evil: "Had some strange poisonous germ crept from body to body till it had reached his own?" (CW, 112–13).[181]

Dorian concludes that "the soul is a terrible reality. It can be bought, and sold, and bartered away. It can be poisoned, or made perfect. There is a soul in each one of us. I know it" (CW, 161).[182]

Dorian has other, better, racial and cultural inheritances: his imagination and Basil's painting, which act as his conscience and spur him to reform. Rejecting egoistic ethics, he says, "I wish I could love. . . . I am too much concentrated on myself. My own personality has become a burden to me" (CW, 154). It was a manifestation of spirit, the imagination, "that set remorse to dog the feet of sin. It was the imagination that made each crime bear its misshapen brood. In the common world of fact the wicked were not punished, nor the good rewarded. Success was given to the strong, failure thrust upon the weak. That was all" (CW, 151). If imagination is the biological and subjective aspect of conscience, Basil's painting is its social and objective representative. Wilde's use of the painting follows precisely the Hegelian conception that art is the spiritual expression of the collective social self. Early in the novel Dorian thinks "the picture, changed or unchanged, would be to him the visible emblem of conscience" (CW, 79). Later, the painting has an obvious moral-aesthetic effect on Dorian that corresponds with Hegel's definition of art's spiritual truth: "*The sensuous and the spiritual which struggle as opposites in the common understanding are revealed as reconciled in the truth expressed by art*. That points us toward what we must vindicate as a higher standpoint for art. Against the view that art is a means of instruction and moral improvement, aspiring to something that remains an *ought to be*, we must maintain rather that its purpose is to reveal *the truth* in an arresting sensuous form, representing for mind the reconciliation of opposites just described" (Paolucci, *Hegel*, 5–6).[183]

In his criticism Wilde always insisted upon an improving, spiritual dimension of true aesthetic creations and response, but his position has been open to misunderstanding because he opposed didactic art per se. Here again Hegel may provide some light: "That art has taught, that it has been in fact the first *instructress* of peoples is certainly true. But to suggest that *didactic* utility in the moral or spiritual sphere is its end and that its sensuous basis is a mere means, amounts to denying it a vocation and purpose of its own. The sensuous and the spiritual are both essential to art" (Paolucci, 5). The painting shows Dorian's spiritual degradation in sensuous form, and Dorian perceives his immorality as ugliness, the perfect representation of the spirit in art. His reaction to the painting is precisely the sort of synthetic response Hegel expected: It moves and disturbs him. However, when he describes the power of art, Lord Henry denies it: "Art has no influence upon action. It annihilates the desire to act. It is superbly sterile" (CW, 163). This response is part of an extended conversation near the novel's end which ironically reverses the roles of instructor and innocent Lord Henry and Dorian play in their first meeting. Then Dorian was the innocent. Here Lord Henry cannot believe that Dorian has committed a crime, denies that he has poisoned Dorian with a book, or that the soul with all its power exists.

Lord Henry had identified conscience as mere cowardice early in the novel, but the ending of the book denies this skeptical view. Dorian's conscience continues to bother him even though, with the death of James Vane, "he knew he was safe"

(CW, 157). In fact, just before he stabs the painting he again reflects that he has no practical fear of punishment for his crimes: "James Vane was hidden in a nameless grave in Selby Churchyard. Alan Campbell had shot himself one night in his laboratory, but had not revealed the secret that he had been forced to know. The excitement, such as it was, over Basil Hallward's disappearance would soon pass away. It was already waning. He was perfectly safe there" (CW, 165). Wilde's design rules out Lord Henry's shallow identification of conscience with cowardice. "It was," thinks Dorian, "the living death of his own soul that troubled him" (CW, 165). The conscience which moves him is motivated not by fear but by his desire to be good and to restore beauty to the painting (that is, by moral-aesthetic feelings): "Perhaps if his life became pure, he would be able to expel every sign of evil passion from the face" (CW, 165).

If Dorian's conscience is not motivated by cowardice, neither is it quieted by his conventional good deed, his refusal to seduce Hetty Merton. The painting as conscience interprets this act as hypocrisy because Dorian does it in the spirit of calculation rather than love. With this episode we believe Wilde underlines his idealist belief that the good is a mental attitude rather than an evaluation of the consequences of acts.

Dorian's self-destruction comes, fittingly, through the interaction of his imagination and the portrait mirroring the true state of his soul. He considers two possible responses to changes in the painting: He can confess and take responsibility for the past or destroy the last remaining evidence and free himself from his "monstrous soul-life" (CW, 167). The first—the duty to confess, to suffer public shame and make atonement—suggests Dorian can free himself only by satisfying an inherited representative of society in his psyche. Such a psyche is like Clifford's description of the individual ego as a later development than the powerful tribal self. But Dorian rejects the course of conscience and chooses to disguise his crime. He acts on the reductive materialist conception of the psyche as ego, rather than the idealist one he had just stated to Lord Henry, and so rejects a true contrition. He returns to Lord Henry's suggestion that his renunciation of Hetty represented nothing more than vanity: "Had there been nothing more in his renunciation than that? There had been something more. At least he thought so. But who could tell? . . . No. There had been nothing more. Through vanity he had spared her" (CW, 166).

After this soul struggle he decides to destroy the painting because it no longer gives him pleasure (a purely hedonistic rationale); though he recognizes the need for soul, he cannot free himself of egotism and admit his moral error. In short, he never changes his mind. Such confusion is required for the great dramatic stroke of the conclusion. Dorian must stab the portrait: "It would kill the past and when that was dead he would be free. It would kill this monstrous soul-life and . . . he would be at peace" (CW, 167). All this comes true, but not as Dorian had planned. His attack on his symbolic self is turned by the more fundamental tribal self to an attack on his physical being. To kill the past is to kill the self. The spiritual and social self is stronger, the action suggests, than the individual, self-regarding reason. Dorian and Basil, the representatives of the imaginative race, see both the letter and the spirit, the surface and the symbol, in Dorian's portrait. In this context Wilde's comment on the "terrible moral" has a precise and understandable meaning: "Yes, there is a terrible moral in *Dorian Gray*—a moral which the prurient will not be able to see in it, but which will be revealed to all whose minds are healthy" (*Letters*, 259).

While *Dorian Gray* has the attractions of a melodramatic plot and fascinatingly sinful characters, we find it finally less satisfying as "art-criticism" than "The Portrait of Mr. W. H," which has an intellectual expansiveness and resilience lacking in the novel. To our tastes "W.H." more elegantly and completely fuses Wilde's systematic intellectual training and his narrative artistry. We imagine Wilde might have developed as a writer of philosophical "art-criticism" as well as a dramatist had he not foundered on his own self-destructive impulses and the wrath of the Philistine public.

However, *Dorian Gray*, Wilde's last effort at "art-criticism," shows the continuing importance of the synthesis of philosophical, scientific, and aesthetic ideas he first developed in the notebooks and expressed in the "Rise." In tracing the influence of the synthesis we have stressed the coherence and clarity of both, since prevailing opinion paints a portrait of Wilde as an eclectic wit and prophet of modernism who played with ideas but did not believe in them. Perhaps this is only to say that critics with modernist tastes ignored both the origins of Wilde's thought in nineteenth-century theories and the unfashionable sensibility in much of his poetry, fairy tales, costume drama, and *De Profundis*. Such misprision may have been necessary to forge a modernist tradition, but in the longer tradition of historical scholarship forgeries are generally discovered. The Wilde we find in the notebooks brings us closer to a historical understanding of a complex intellect and to the ideas which unify and explain much in his critical and fictional work. We think we have painted a faithful portrait of Wilde as the last Victorian sage. Sometimes, when we contemplate it, we believe we have realized his intentions.

Utopian Conclusion

Our judgment of Wilde's critical theory, its relevance and importance, is implicit in the critical practice that has shaped this study. Like Wordsworth, we find in stones the sermons we have hidden there: We value the scholarly and critical techniques which help to recover, however imperfectly, the historical meanings of texts. What value do such meanings have, aside from their revisionary understanding of Wilde? How does revaluation of Wilde contribute to the debate Arnold began about the function of criticism? We suggest that Wilde offers corroboration of the importance of critical method, and that his relevance can be suggested emblematically in relation to two representative figures, Arnold himself and one of our modern practitioners, Edward Said.

Wilde's revision of Arnold, his effort to define cultural authority as a process resisting the inertia of the dominant culture, remains a crucial task for the institution of criticism. Wilde attempted to take criticism out of the wilderness where Arnold had left it (CB, 55). Unlike Arnold, Wilde could not endorse Joubert's dictum of "Force till right is ready" and its attendant state of critical authority.

Such a position left the critic in the contradictory situation of simultaneously questioning and authorizing monolithic, "natural" culture based on the perception of "right reason." Wilde tried to revise Arnold and to confront these social contradictions; he took many of Arnold's premises to their theoretical and utopian conclusions and advocated oppositional engagement leading to political and individual self-realization. Furthermore, unlike Arnold, Wilde developed his theoretical position out of a foundation in philosophy and nineteenth-century evolutionary theory. Where Arnold mystified or ironized his own theoretical assumptions, Wilde developed a coherent idealist-socialist position which gave the critic a powerful dialectical method to enact an oppositional discourse, to shift the Philistine culture from its inertial acceptance of pre-digested, "natural," dominant culture.

Today the function of criticism in the transmission and revision of culture remains difficult and problematic. For instance, Edward Said, in "Secular Criticism," describes modern cultural history as a "three-part process" which, in spite of its impetus toward the creation of "compensatory order," "can easily become a system of thought no less orthodox than culture itself."[184] Wilde, however, recognized the constant urge toward stability and the need for a continual dialectical advance and revision. Said might object that Wilde's progressive, revisionary dialectic is itself the product or function of a totalizing system, the Hegelian synthesis Wilde developed in his notebooks. Said's "secular criticism" must project itself upon the world and the text without any assumption of its correspondence with an idealist world order. However, removing the totalizing force of Hegelian idealism from Wilde's system does not invalidate the power of dialectical thought in intellectual inquiry, as Said's description of the "three-part pattern" in modern cultural history demonstrates. Wilde's essential contribution to the present is not the anticipation of any particular critical system (as critics from the beginning of this century to the present have claimed) but rather a basic position about the necessity of a dialectical view which complicates and enriches the production and reception of art and criticism and recognizes its place and relation to other institutions. In proposing that "a Truth in art is that whose contradictory is also true" (CW, 1078) and that "progress in thought is the assertion of individualism against authority, and progress in matter is the differentiation and specialization of function: those organisms which are entirely subject to external influences do not progress any more than a mind entirely subject to authority" (CB, 69), Wilde stated fundamental assumptions about the function of criticism at *any* time. As Said puts it for the 1980s, "[C]riticism belongs in that potential space inside civil society, acting on behalf of those alternative acts and alternative intentions whose advancement is a fundamental human and intellectual obligation" (WTC, 29–30).

Wilde's intentions, we contend, can be understood and used without their totalizing system to produce the kind of oppositional, ironic criticism Said argues for: "[I]n the main . . . criticism must think of itself as life-enhancing and constitutively opposed to every form of tyranny, domination, and abuse; its social goals are noncoercive knowledge produced in the interests of human freedom" (WTC, 29). Its method must rely not upon some essentialized warrant of the mind's search for its own self-perfection or the realization of the Idea in History; we must leave behind such nineteenth-century external criteria for the authorizing of progress and freedom. Rather, we must select from Wilde's system what he described in the notebooks as "the dialectical energy of thought . . . the spirit of scepticism—the force of contradiction and the negative" (CB, 205) to empower us: "A map of the world that does not include Utopia is not worth even glancing at, for it leaves out the one country at which Humanity is always landing. And when Humanity lands there, it looks out, and seeing a better country, sets sail. Progress is the realisation of Utopias" (CW, 1089).

4

Oscar Wilde, Commonplace Book

* * *

[1]

Commonplace Book•

Metre

________________•

the inadequacy of the xviii cent. English Literature due to the use of rhymed couplets: cf. also Drydens tragedies: so Moliere is really best in his prose not his verse•

The limitation, that is, of the rhymed alexandrine, is too narrow• all art of course works with limitations and only reaches its perfection by the adequate recognition of its limitations, but the limitations must be rational: rhymed alexandrines are as trammelling as the use of monochrome to a painter, or the chisel alone to the sculptor• We must not however underrate the value of Racine and his school in forming the French genius: They were important to the French: as S. Beauve said of Lamartine "He was important to us"

* * *

[2]

Thought

A wave of opinion reaching a certain height cannot be changed by any evidence or argument• but has to spend itself in the gradual course of things before a reaction takes place•

:+: so there is nothing in Moliere to compare with the august gladness and assurance of progress contained in Periklēs of Tyre•

Nonconformists•

To Dissenters we owe in England Robinson Crusoe• Pilgrims Progress: Milton: Matthew Arnold is unjust to them because "not to conform to what is established" is merely a synonym for progress•

* * *

[3]

Culture•

The human spirit cannot live right if it lives by one point alone: it has a vital need for conduct and religion but also for beauty• social life• intellect• manners &c.

Tragedy•

tragedy breasts the pressures of life• comedy eludes them by

:+: its irony• but the ultimate peace and assurance rests with tragedy• that is nothing ever comes of shirking a problem• the answer obtained by the over simplification of a problem is only approximately valuable•

Whistler•

We cannot forego tears and laughter for a scientific adjustment of different coloured blues• technique is the alphabet of art•

* * *

[4]

[This paragraph appears opposite the second paragraph on page 5.]

Prevision• in social life of course prevision is not always accurate but where there is some prevision there is some science as in Meteorology• all science is not exact science• and as regards the obj• that history never repeats itself and so that we cannot test the truth of our hypotheses, in no concrete science is there absolute repetition• Even in astronomy the repetitions are only approximate•

* * *

[5]

Sociology•

Is it possible to ascertain the laws which govern the phenomena of human action, the complex manifestations of that redistribution of matter and force which we call Society: the objections to the science may be summed up as those resting on the assumption of "extra-natural interference", of a "certain causeless spontaneity called the Will", and the "Great man theory of Life."

Even however by these it is generally admitted that social changes may be produced by natural causes[,] that after certain antecedents certain consequents will be sure to happen, in a word that prevision is possible and where prevision is possible there is science[.]

Method•

Every aggregate of units is determined by the properties of its units• man is the unit of society• so to understand social action, it is necessary to understand

✵ ✵ ✵

[7]

human nature• there can be no knowledge of human nature without knowledge of the Laws of Mind, (Psychology) nor of the Laws of Mind without knowledge of the Laws of Life (Biology)•

The science of society then rests on the science of life: sociology on Biology• the state as Plato saw is "man writ large": now in spite of his imperfect psychology, and ignorance of biology Plato saw the real analogy between the individual and the social organism•

For what characterises an organism as opposed to the undifferentiated aggregate of protoplasm, is the mutual dependance of parts, the increased differentiation of function and of structure• The "physiological division of labour" which Milne Edwards showed was the active law of the individual organism, corresponds to the Politico Economical principle of the division of Labour recognised by Plato as the law of the social organism•

✵ ✵ ✵

[9]

So in the history of society we first have a chaotic aggregate of human beings corresponding to the formless protoplasm• In the primitive life there is no mutual dependance, no exchange of services, as in the rudimentary animal there is no vascular system• The developement of the nervous system and the differentiation of function in the individual organism corresponds to increased interdependance and the division of Labour in the social organism•

✻ ✻ ✻

[13]

X X X The Sciences in their Educational aspect as forming habits of mind.

From the abstract sciences we gain the perception of the necessity of relation, the absoluten~~ess of u~~niformities: it's danger is that mathematicians are apt to simplify their problems too much and to be bad reasoners on contingent matter•

From the (abstract) (concrete) sciences we gain conception of cause and effect• it's danger is that it makes us too analytical•

From the concrete sciences we gain ideas of continuity• complexity• and contingency• and from the concrete• organic• sciences that of the fructification of causes•

Abstract sciences (as Logic• Geometry) give us the forms of phenomena•
Abstract-concrete• sciences• (as Physics• Chemistry)• give us the factors of phenomena•
Concrete sciences give us the products in Their synthetical totality•

✻ ✻ ✻

[14]

Life is coextensive with protoplasm, but is consciousness coextensive with life•?

Progress.

is simply the instinct of self-preservation in humanity, the desire to affirm one's own essence• knowledge rendered active by emotion of self-preservation produces higher forms of civilization[.]

Mankind has been continually entering the prisons of Puritanism, Philistinism, Sensualism, Fanaticism, and turning the key on its own spirit: but after a time there is an enormous desire for higher freedom—for self-preservation.

✻ ✻ ✻

[15]

In Sociology as well as in physical science experiments we must make allowance for the "personal equation" and for the 'refraction' due to the historic medium•

Matter often behaves paradoxically and so does mind: that two cold liquids mixed together should become boiling hot, is as strange as the fact that severe laws do <u>not</u> decrease crime•

Light falling on a crystal alters its molecular arrangements, but can only do so by an infinite series of millions of etherial waves, so also in the social organism any permanent change requires innumerable recurrences of thoughts• feelings• actions•

X In the bodily as in the social organism the normal principle is one of compromise• [The X is penciled in the margin.]

* * *

[16]

Unity of the vital principle of plants and animals shown by (1) the identification of "sarcode" with vegetable protoplasm or "chlorophyll". (2) Claude Bernards experiments that plants can be put under the influence of anaesthetics: and also the fact that yeast in water respires and breathes like an aquatic animal. also Fly catching plants—Respiration was long supposed to be a barrier between the animal and vegetable life but the yeast experiments break that down[.]

* * *

[17]

"Combien de ceux qui nient' l'immortalite meriteraient une belle deception"

<u>Renan</u>•

<u>the Unity of the Principle of Life</u>

The structureless semi•fluid substance called "<u>Sarcode</u>", which is the basis of the bodies of the lowest members of the animal kingdom, was discovered by Schultze to be identical with the "protoplasm" of plants: this protoplasm is the constant basis of the varying phenomena of life• and its own attribute is "<u>irritability</u>"

The problem of science is "consciousness[.]" Is thought a property of the cerebral protoplasm in the same sense as irritability is? It is contended that as physiological activity is the property of every living cell, psychical activity must be so too and a "cell-soul" has been invented as a conception inseparable from that of life•

To point out where [continued on p. 19]

* * *

[18]

X

The Protoplasmic Hierarchy•

I. Bathybius Haeckelii: structureless albuminoid mass.
II. The Amoeba: here we find the "cell", the morphological unit of organization, the physiological source of specialised function and individualization: but there are unicellular organizations—we rise To
III[.] multicellular organizations where every cell has it's own special work and yet all combine for the welfare and unity of the whole:
IV[.] where to the attribute of irritability common to all protoplasmic matter is added that of conscious volition: at a certain period we are met by The impassable barrier of thought[.]

* * *

[19]

[continued from p. 17] consciousness begins in the Life Kingdom is extremely difficult• much that seems the result of volition (as the movement of algae.) are acts absolutely unconscious: upon the other hand the limitation of our faculties prevents us conceiving any substance different from that of matter: but these limitations are subjective:

As regards the hierarchy of protoplasmic phenomena we can ascend gradually by increased differentiation of function and division of labour from the structureless albumenoid matter (Bathybius Haeckelii) which the depths of the north Ocean hide, to the elaborate cerebral cells of the human brain which if they are not themselves consciousness are at least the organs by which consciousness manifests itself:

* * *

[21]

Literary Criticism among the Ancients•

Criticism at first appears as a habit of mind, an attitude of public opinion, rather than a formulated system• however the first regular literary critics were the judges appointed to decide between the rival rhapsodists: an institution which continued from the early contests of the Homeridae in Chios, and the literary reunions of

Hesiodic Boeotia, over to The union of lyric and epic poetry in the Dionysiac Theatre of Athens•

So far for the formal institution of literary critics, as for literary criticism itself it first appears in Greek literature wearing the mask of Comedy: and this seems to have been in a manner natural, for while tragedy always preserved the limitations which its epic and religious origin imposed on it, comedy from its very essence became

* * *

[23]

the mouthpiece of public opinion: In Sicily however comedy did not have that peculiar literary aspect which she acquired in Athens• There is a piece of Epicharmos' entitled "Herakleitos" and the sophists are laughed at in the Mimes of Sophron, but as a whole the shafts of the Dorian comedy were not aimed at literature•

At Athens before Aristophanes we have parodies of plays and legends brought out wh. we may regard as a rude form of criticism, and Pherecrates seems to have largely criticised his contemporaries, such as the tragic poet Melampos, and the "Music of the Future" is interpreted by Cinesias and Timotheus, but the Frogs of Aristophanes form an era both in the developement of comedy, and the history of literary criticism•

Euripides, who was criticised by the conservatives of his own
X day much as Swinburne is by the Philistines of ours, is there attacked for the laxity and extravagance of his
[A large X in pencil appears in the margin.]

* * *

[25]

metres as well as for the immorality of his teaching: and there are in the main the two channels into which the stream of Greek literary criticism flowed—poets were criticised on ethical grounds, and on the grounds of style•

The criticism of the poets on ethical grounds began with the philosophers such as Xenophanes; metres, style and the like are first treated by the comedians: Plato employs both canons in his criticism on Homer, elsewhere treating of the method and faculty of the poet, and the nature of the beautiful: but true criticism is always ~~deductive~~ inductive; abstract speculations on Τὸ *Κάλον* form a bourne from which no critic ever returns• and Aristotle may rank as the first really literary critic of ancient times:

* * *

[29]

After Aristotle criticism and literary activity generally is concentrated in Alexandria and Pergamos• but the movement was characterised more by erudition than by any real critical spirit: still we must mention 1. Aristarchus• whose criticisms on Homer and the first application of the principle of "l'homme et le milieu" are most important[.]

Rejecting the useless allegorical interpretation[,] he sought to explain the immoral stories of Homer by reference to the age of the poet and the imperfect civilization in which he lived: this acknowledgement of the union between history and the philosophy of art was rare in ancient criticism• it may be found however in Plutarchs parallel lives of Demosthenes and Cicero, and also in Cicero's Brutus: Herodotus and Aristotle and Plato had all pointed out the connexion between the character of a race and the physiological effects of climate and situation; but to carry the same ideas into the lives of individuals was not common•

* * *

[31]

2. Strabo• to whom we owe the important remark that poetry always precedes prose in the historical developement of Literature•

3. Dionysius of Helicarnassus.

Judgement on the orators and historians: enters into questions of style to prove authenticity of writings:

4. Plutarch•

On the beautiful• On poetry• Comparison of Aristophanes & Menander•

5. Dio Chrysostom

very important critic: his "Olympiacus" may be regarded, so far as his comparison of art and poetry is concerned, as the precursor of Lessings' Laocoon: and who has a brilliant passage on the limitations of the sculptor who has to deal merely with form•

6. Lucian: On Dancing especially•

* * *

[33]

Philostratos• to whom we owe the first definition of the imagination which Plato and Aristotle had passed over[.]

Longinus• περὶ υψους•

Criticism went always with Philosophy—To Plato beauty was an intuition: Aristotle analyzed it: Sextus Empiricus denied it: the Stoics took no notice of it: the Epicureans dwelt on its subjective affect: Longinus showed how the sublime had its root in human nature• dwelt on it's psychological effect• and this is exactly what Burke in his Essay on the Sublime and beautiful did: he sought for the principles of art in the right place—in the faculties and sentiments to which art makes it's appeal•

✱ ✱ ✱

[37]

''dans la vie morale il est beau de quelquefois faire naufrage''.
The Italian art had it's root in the mosaic painting of the Greek artists, and the ancient Greek sarcophagus of Nicholas of Pisa: from the one it got it's colour from the other its form• it added itself the pity and passion, the human background to all art.

✱ ✱ ✱

[39]

Punishment as a motive
Whether the dead rise or not it is at least a good thing to have fought with the beasts of unrighteousness at Ephesus or elsewhere• the incapacity of severe criminal legislation to produce higher morality or even order may be parallelled by feeble practical influence exercised on men by the conceptions of heaven and hell: nothing speaks so well for the noble nature of man as his showy indifference to any system of rewards and punishments either heavenly or terrestrial• No man is turned towards the path of righteousness from any fear of a future place of punishment, are though [although?] it be a place where the worm dieth not and the fire is not quenched•

✱ ✱ ✱

[40]

[This entry appears opposite ''Mommsen says . . .'' on p. 41.]
X
Different Education of a Roman and a Greek boy.
The Roman was educated for the family and the state: to be a pater familias and a civis: the refinement of Greek culture coming through the romantic medium of impassioned friendships, the freedom and gladness of the palaestra, were unknown to the boy whose early recollections were those of the senate house and the farm. So [to] the one gravitas was the ideal, to the other ''*ευτραπελια*'':
The two Greek ideals
''*φιλοσοφειν μετ' ερφτος*
φιλοκαλειν μετ' ευτελειας''
had no place in the narrow conservative education of a race who gave up art to their slaves and made it a handicraft and expelled the philosopher from Rome• and lastly Greece represented individualism—Rome citizenship[.]

✱ ✱ ✱

[41]

Roman literature began with a discussion on the theory of agriculture: the early Greek poems have for their subject maritime adventure and war• Shows spirit of each nation.
Mommsen says 'the beginnings of poetry belong rather to women than to men: so the Camenae and the Muses were feminine'': is this so?
Romans had no native word for a poet. poeta is Greek: vates is Keltic:
The centuries during which Rome possessed no literature are those of her real greatness: when their literature did appear it by no means represented the spontaneous working of the Roman spirit but was imported as a necessary luxury from abroad: Roman life lacked that joyousness, and freedom, without which art is impossible—Their first writer was a prisoner of war, a slave, and a Sicilian Greek• Their first introduction to philosophy was through the shallow rationalism of Euhemeros, to dramatic art through Euripides—

* * *

[43]

Appius Claudius Caecus•

Between the Tarquins and the Caesars, that long five hundred years of extraordinary deeds and ordinary men Appius Claudius Caecus is the great link: he was the incarnation of the idea of progress, neither an aristocrat or a democrat, but rather a political humanist.

From him date the Roman Roads, the network of the world: Roman jurisprudence under whose aegis we still live: he saw how Romes mission was material civilization, and organization and order: from him too date poetry• eloquence• and grammar•

The attempt of Pyrrhus to found a material sovereignty of Hellas in the West was crushed by his hand: but he represented the Greek ideal of individualization and paved the road• for the intellectual supremacy of Hellenic thought•

Romes greatest maxim that the ''Republic negotiates not while there are foreign troops on Italian soil'' came from him• but he came either too early or too late• the Gods made him blind on account of his untimely wisdom•

* * *

[45]

Individualism in private life was checked by the censors—the champions of monotonous uniformity[,] the incarnations of the Spartan ideal—while individualism in public life was held back by the short Tenure and limitations of office, by the many checks afforded by the constitution, and by the strong class antipathies—party feeling being always more excited for a class than a person: when Romes Empire did at last fall under the dominion of one man it was found necessary to surround him with an atmosphere of divinity and to claim for him parentage different from that of ordinary men: The one man who did on the principle of individualism, and recline [reclaimed?] the Periklean ideal of the "Government by the best man" was assassinated before he had worn the purple a year: his successors were wiser and "hedged the[ms]el[ve]s with divinity"

* * *

[47]

The history of every nation and of Italian history in particular is *Συνοικὶσμος* on a grand scale: the passage from cantonal individuality to national unity: the growth of an organism in fact: *Συνοικὶσμος* is in Politics, what the evolution of multicellular organisms from unicellular (such as amoebae) is in the sphere of science: The division of Labour, the differentiation of function, the evolution of organisms can be illustrated from History as clearly as from the microscope.

History is the account of the mutual attraction and repulsion of primitive political atoms.

The atomic theory is a valueable hypothesis not merely in physics but in politics as well.

* * *

[49]

<u>Law•</u>

Law has it's root in custom, custom in the psychological principle of imitation and dependance on others: Individualism, or the attempt to affirm one's own essence, is a late product of matured civilization•

The test of good law is the recognition that the ultimate sanction of it's ordinances is to [be] found in reason• Rationalism working in the sphere of custom produces Law:

Roman history comes before us in an exceedingly matured state: from the first it exhibits a comparatively modern character: there are in it fewer traces of the primitive condition of things than in any other Indo•Germanic race• There are no traces of ordeal by combat, or of the blood-guiltiness of a <u>family</u> consequent on the crime of an <u>individual</u>, or of the influence which the comrades of the accused were entitled to exercise over the pronouncing of judgement, all characteristics of Germanic primitive life: The abolition of the right of blood revenge in early times is shown by the primitive myth of Tatius, as the [continued on p. 51]

* * *

[50]

› Just as they set aside allegory and personification in their religion, so they expelled symbolism from their law.

› When Vergil in his apostrophe to Rome as the predestined mistress of the World says alii . . . orabunt causas melius . . He must refer to eloquence not to legal knowledge : Romes mission was "dare jura"

X so Terence thinks a play is damned because it has a mistake in Law.

* * *

[51]

[continued from p. 49] early right of provocatio is in the myth of the Horatii:

Roman legislation was great because it never allowed itself to be trammelled by symbols—what it kept in view always was the motive and intention of the parties engaged—never except in the case of certain religious ceremonies such as confarreatio did it allow to symbols that objective value which is the characteristic of irrational legislation and immature civilization: it also provided a permanent board to remedy the laws and keep them in accordance with the requisitions of the age:

Jurisprudence is the only part of literature which was developed by the Ro-

\+ mans in a thoroughly national spirit and the more national a poet is the greater place does Law find in his writings as in Plautus:

Invested at first with a theological and religious sanction, and confined to a narrow oligarchy, it was made public and consequently rational by that great humanist and [continued on p. 53]

* * *

[52]

Cicero De Or. I. 45. calls attention to the splendid position of Roman "jurisconsulti" as opposed to the Greek *πραγματικοι*• "Domus jurisconsulti totum oraculum civitatis"

Catus. "egregie cordatus homo catus AEliu' Sextus"

In early times Law was kept popular through it's connection with politics and the study of eloquence afterwards under the Empire when eloquence and political ambition were both silenced it became more scientific—specialised and differentiated•

Quinctilian talks of the charter lawyers who give themselves up "ad album et rubricas" : and a tombstone has "HUIC MONUMENTO DOLUS MALUS ABESTO ET JURISCONSULTUS"

* * *

[53]

[continued from p. 51] pioneer of cosmopolitanism Appius Claudius Caecus: Catus was the first juridical writer : the "responsa" or "auctoritas pendentum" went

as in English Law side by side with the actual edicts: R. Influence of the Praetor's Edict•
—of Aristotle and the Stoical Philosophy• as ref. jus naturale and *φύσει δικαίον*•
—Appointment of the Praetor Peregrinus first beginning of Romes Empire.
—How Law suited the positivist materialistic notions of the Romans
—Dare jura• synonym for Empire.

The greatness of Rome rested on the fact that the Roman people endured and ordained a system of Law where the eternal principles of liberty and subordination, of property and rightful redress reigned and still reign unadulterated and unmodified• And though the root of the Christian doctrine of the atonement ~~had its r~~ is to be found in the scapegoat of the wilderness, yet in it's intellectual progression it undoubtedly received much assistance from the analogy afforded by Roman Law, as the trinity and the Logos were expanded by the Greek mystics—

✱ ✱ ✱

[55]

Poetry

In proportion as poetry separates itself from human passions and feeling, so does it's [*sic*] lose its own essence, and the quality of it's power• Wordsworths sonnet on the advantage of Compulsory Education is as unfit a subject for poetic art as are those flights of transcendent imagination to which Shelley sometimes soared ~~fell~~•

One flies too high: the other does not fly at all: So the pure intellect and the pure imagination form not themselves the right mainsprings of noble song which has it's natural roots in the passionate side of nature:

(now it would seem the function of criticism to present the right materials to the poet• Criticism prepares the way for the creative faculty, and like the Baptist its voice often cries in the Wilderness:)•

So the Feminine nature ascribed to the Muses—though there may be historical considerations of primitive manners to account for it—will at any rate symbolise for us the close connection of poetry with Feeling and sentiment—and even when Apollo took the Place of the [continued on p. 57]

✱ ✱ ✱

[56]

Euripides•

When we criticise Euripides we must remember that it is not the highest works of the imagination which are the most affecting•

✱ ✱ ✱

[57]

[continued from p. 55] Goddesses of Song his status always preserved some of that effeminacy and languor and voluptuousness which are the characteristics of that "passionate humanity" which is the background of true poetry•

✷ ✷ ✷

[59]

English Thought from Bacon•

With the introduction of the inductive method of research, the rise of democracy and the experimental sciences, there arose in England a reaction against "innate ideas", the palladium of conservatism and the deductive method: The problem before men was what is the Origin of Knowledge• Locke found this in sensationalism and the reality of an external World: Berkeley fell back on Spirit or God as the fountain head of Knowledge• but Hume the logical outcome of these two thinkers, and the consistent sceptic saw that apart from Sensation matter and mind are equally illusions:

Hume is the champion of Sensationalism and Empiricism: The common sense of mankind revolting against the attempted suicide of Philosophy and the annihilation of science found its defenders in the uncritical Scotch school which in the developement of thought occupies no position at all: The line of dialectic was carried on by Kant and Spencer: the first analysed sensationalism into the three component parts of sensation[,] perception and Thought, reinforced the innateness of ideas in the forms or categories of

✷ ✷ ✷

[61]

thought, while Spencer substituted for the individualistic empiricism, the generalistic empiricism of hereditary transmission of concepts: Innate Ideas have thus returned to the mind, in Kant on transcendental, in Spencer on Biological Grounds:

To these notions of the pendulum of thought and developement of the dialectical idea Mill cannot be said to have contributed anything:

✷ ✷ ✷

[63]

Slavery•

Roman law from the first denied to the slave any legal rights: the Greeks on the contrary recognised the marriage of slaves as a legal relation: It is remarkable that the two occasions when the rise of a feeling of humanity towards slaves in Roman history can be noticed—after the fall of the Republic and after the introduction of

Christianity, are both accompanied by increased severity of legislation: under Augustus and Constantine.

Notwithstanding that satirists inveighed agst [against?] the cruel treatment of slaves, that philosophers claimed for them the natural rights of humanity, and that lawyers even pronounced slavery to be 'contra leges naturae', still no ancient writer ever seems to have contemplated its overthrow•

It was to them a necessary institution of Political Economy: theoretically it was an injustice • practically a necessity.

* * *

[65]

Mind and Body•

Just as Aristotle's theory of habit may be said to have given Ethics a physical basis, so in our own day we have a physical basis of mind: Henry More said That [The rest of the page is tantalizingly blank.]

* * *

[68]

[This entry appears opposite the first entry on p. 69.]
So Lucretius says that if men had not kept their words to one another the whole race would have been cut off•

* * *

[69]

Survival of Fittest in thought

Nature kills off all those who do not believe in the Uniformity of Nature and the Law of Causation.

Progress

Progress in thought is the assertion of individualism against authority, and progress in matter is the differentiation and specialization of function: those organisms which are entirely subject to external influences do not progress any more than a mind entirely subject to authority:
The Love which Plato regarded as the essential concomitant of philosophy was the Love of the beautiful—a love both abstract and concrete—of *το καλον* as well as of Charmides—but Spinoza's love is an amor intellectualis Dei and the modern love is the love of truth for truth's sake•

* * *

[71]

Agriculture

Among the Italians as among the Hellenes all the institutions of life, the legends[,] laws and manners are associated with agriculture: R. how agriculture is connected with marriage and the founding of cities: Ceres. Tellus. Demeter preside over it. the oldest form of Roman marriage is "confarreatio" *ἄροτος* was the Gk. begetting of children: in the founding of cities the plough was used.

Philosophy

S• Augustine rose up from the Hortensius of Cicero with the words 'surgere coeperam ut ad te redirem"—Lord! I return to thee! and says that Philosophy descried "de silvestri cacumine patriam pacis", the road to which Christianity held•

* * *

[72]

[The entries on this page appear opposite the remarks on Roman religion recorded on p. 73.]
Gods of the Indigitamenta.
(For childhood)
Vaticanus
Fabulinus
Educa.
Polina.
Caba.
Abeona
Herduca. [Iterduca?]
Religion like Socratic Philosophy was brought down from the heavens to earth: the commonest incidents of a commonplace life had their special deities: The Romans would have regarded a temple '*Αγνωστω Θεω*' as a reflection on the legal-religious knowledge of their pontiffs: when the great Carthaginian retreated from before Rome they had no doubt whatever about the special intervention of Rediculus Taditanus and the first word a baby uttered was one to particular grace of Fabulinus—

* * *

[73]

Roman Religion.

The reason for the low position held by priests in the Roman Religion was the legal principle that pleading could not be done by deputy: the Relation of a Roman to his God partook of all the formality and the mutual obligations of a civil contract. The Roman religion mirrors the four great qualities of the Roman people[:] Positivism• Patriotism• Domesticity• Law•

They had not the creative imaginative power of the Greeks to give life to the dry bones of their abstractions: they had no art: no myths: For 170 years says Varro there were no statues at Rome• and the most ancient and native Goddess—Vesta—the one Goddess common to [blank space] preserved to the end her abstract nature and was symbolised by a blazing fire•

The utter want of productive power in the Roman Religion was literine [literally?] the ultimate cause of the thorough poverty which marked Roman poetry and Roman speculation[.]

* * *

[74]

+ the normal conditions of progress in thought is this: first a narrow definiteness, an uncompromising dogmatism• <u>then</u> the antagonism and criticism to which this gives rise• lastly the intellectual synthesis and union: (Hegel) now the first condition was not observed in the Roman religion—

* * *

[75]

The Greek spirit prepared the way for the wisdom of it's matured manhood, by the free• imagination of its youth: Philosophy must come as a reaction: now the Roman religion was eminently respectable—in the alliance between Church and State there was nothing to react against: Like modern Protestantism it concerned itself primarily with conduct, and represented the statical power of their civilization: the claims it made on reason and faith were of the slightest possible• Lucretius was obliged to go to the Greek religion to find an object of attack: When he announced that Ceres and Liber were merely abstractions for Bread and Wine the Roman spirit received no shock—they had never been anything else—and he was forced to go to Greek legend to instance his line "Quantum religio potest suadere malorum"

* * *

[76]

The Renaissance as an intellectual movement in which speculation was of more value than conduct:

The first post-republican centuries as an emotional movement in which philosophy and science were of little importance, where the 'tauribolium' and the purificatory rites to Isis were of more value than abstract speculation[.]

* * *

[77]

Renaissance and the Empire

The sin and superstition of the lonely Emperor at Capri, Nero's terrible joy in all things evil and beautiful, Vitellius more of an animal than a man, Domitian with his coward hatred of all good men and aesthetic sensuality—to parallel these things we must go to the Italy of the Renaissance, to the Baglioni and Sfozzes [*sic*]—to Malatesta at Rimini—to the nameless horror of the Vatican of the cinquecento•

Yet as far as the Renaissance is concerned what humanity demanded at that splendid crisis, when the world-spirit was in the throes of travail of things evil and good—was not new theories of conduct or even its practice—not order—not virtue even—but free scope for the intellect—the throwing off of authority to breathe again in the free frank air where nothing stood between men's eyes and the sun of truth: the Emotions had had their day—the intellect now asserted it's right: The Renaissance was served better by Machiavelli than by Savonarola, by Cellini than by Francis.

✳ ✳ ✳

[79]

But while the Renaissance was an intellectual movement, the Christian religion required as it's precursor no philosophy or art but an emotional and spiritual awakening—a desire for purification—a hatred of sin—a love of God—all which things came as a reaction against the horrors of civil war first—and the horrors of Roman society afterwards—

✳ ✳ ✳

[81]

Ideals.

Nothing is more pathetic than the fruitless attempts to present high moral ideals to a people who have not yet fulfilled the requisite physical qualifications for their reception•

Science and Religion

Primitive religions contain the germ of philosophy and of physical science—unnatural children who seek to annihilate their mother when they have attained to their maturity• yet the intellectual synthesis between religion and science as it is commonly given to us is nothing more than an monstrous OEdipodean union of vigorous manhood with the effete mother who bare it—a union whose children must be wanderers and born to evil things.

✳ ✳ ✳

[83]

Just as the human embryo passes through all the stages of evolution from the lowest to the highest organism during the progress of its perfection, so the human mind must pass through all the stages which have been already taken in the progress of the intellectual world: neither in the world of thought or in that of matter is the past ever annihilated: progress in both must be made by slowly graduated stages from simple sensations, and formless protoplasms, to the highest differentiated organism and the purest abstractions of thought:

✳ ✳ ✳

[85]

The greater permanence of the Roman over the Greek developement, was that Rome was the city of mediocrities: The steady progress of the Spirit of Time is really disturbed and forced into unripe maturity by great men—The light of an Aufklarung is too dazzling for humanity to walk long in: In the developement of man as well as in that of the individual organism those movements are most healthy, and permanent in their result which are unconsciously performed: True progress is the gradual adaptation of means to ends[,] not the conscious straining after a result: Great men are great but the Spirit of Time is greater, that absolute mind which is neither religious or irreligious but infinite and absolute in it's season[.]

✳ ✳ ✳

[87]

The splendour and grace of swift limbs, the grave beauty of girlish foreheads, the physical ecstasy of sensuous life—do we love these things less because the germ of man is to be found in the formless protoplasm of the deep seas, or in the hideous sluggishness of the Lower Amoebae—

as in the physical so it is in the moral life—we turn our eyes not to the deeper depths from which we may have sprung, but to the higher heights to which we can rise•

✳ ✳ ✳

[89]

Modern Positivists are as men who while they deny the existence of the sun yet worship the sunlight on the Earth, who acknowledge that the fruit is sweet to eat, and the flower goodly for sight yet insist that the root is rotten, and the soil barren—

✳ ✳ ✳

[91]

Religions it is said like the sun are most brilliant before death: but with ours this is not so: for the Terrible war against religion of our own day has come not with that clamour and passion which heralds the advent of a new God, not with that persecution which quickens into spasmodic life the exhausted forces of an effete creed and makes Theodora nobler than Cyril, but calmly and steadily from the hell of the philosopher and the laboratory of the man of science:

* * *

[93]

Limits of the Investigation of Nature

It is impossible it is said to comprehend the nature of matter and force: we are obliged to postulate as the basis of life an indivisible inert mass—called an atom–from which forces emanate: The second incomprehensible is consciousness even in its lowest form the sensation of desire and aversion•

It is incomprehensible how to a mass of nitrogen, oxygen, carbon, phosphorus and so on it can be otherwise than indifferent how they lie and move:

With regard to life we must hold that at certain period of refrigeration life appeared as a natural result of it's antecedents: that life is merely the arrangements of molecules in a state of more or less stable equilibrium—their mutual attraction and repulsion which is the source of motion must be simply postulated as an attribute inseparable from their nature: And that introduction of a Force extraneous to matter, which is the [This entry continues on p. 95.]

* * *

[94]

"The phantom of Vital Force which formerly reigned over the whole intestinal canal now scarcely knows where to breed disturbance".

* * *

[95]

[This entry is continued from p. 93.] Vitalistic conception of life—the theory in a word that a particle can change its nature whether it be external or internal to an organism—is now refuted by the rational investigation of the phenomena of animal heat:

* * *

[97]

Useless Organs.

Mole's eyes: the appendix vermiformis in man and the external of the human ear• The navicula of the oyster which causes the extraordinary mortality among young oysters[.]

* * *

[100]

[This entry appears opposite p. 101.]
Note also the shallowness of seeing an opposition between Aristotle and Plato as an idealist versus a realist• In modern phraseology Aristotle wd. have been termed a "mystic", and wd. have defined philosophy as a "knowledge of being or of essence" words to which we can hardly attach any meaning[.]

They are both idealists[,] only to one the idea is actual and immanent (Aristotle) to the other it is potential and transcendent•

* * *

[101]

Idealism and Realism•

The opposition between Idealism and Realism is a shallow one belonging to the onesided method of the understanding: Every true philosophy must be both idealist and realist: for without realism a philosophy would be void of substance and matter• without idealism it wd. be void of form and truth: Realism is the assertion of the claims of the ~~univer~~ particular, the detail, the parts: Idealism is the grasp of the whole and the universal• In the rhythm of both the line of dialectic finds its true course of progress. The earliest philosophies were the most abstract and the most empty: the progress of philosophy is from the empty abstraction which is separate from the particulars to the true universal which is the sum of all the particulars and yet above them• to the totality of things and the absolute Idea:

* * *

[105]

Mind and Matter•

The great difficulties of philosophy have their root in the unphilosophical conception of 'mind' and 'matter' as self-subsistent entities: a chasm was made between the Ego and the non-ego[,] the world of thought and the world of things which they tried in vain to bridge over by such doctrines as preestablished harmony, divine interposition and the like: but this onesided differentiation proceeds from the narrow method of the understanding: to the true idealistic and speculative reason the object and the subject, the Ego and the Non-Ego are really one, and ~~the results of a process~~, are harmonised in the "Original synthetic Unity" which is God—the *νοησις νοησεως*—the Absolute Idea.

The mind does not come forth equipped to conquer the world, nor is the world a prey prepared for the spoiler—they are both the result of a process—

* * *

[106]

The road of philosophy cannot be traversed in a dressing gown—it requires the vestments of a high priest: the sanctuary of truth, no more than the kingdom of God, cannot be entered by a guest who has not a wedding garment:

* * *

[107]

<u>The unity as attained by Philosophy</u>•

The analytic force of science shatters the golden chain which binds the world round the feet of God: it is vain for religion to try and reforge them with instruments really borrowed from the sensible world, or for the popular consciousness to assert that the links have never been severed at all, the one power which is able to weld together the isolated fragments into a true synthetic unity is the power of philosophy—and it does so by transporting them into a higher sphere—the sphere of pure thought—where matter and form, mind and nature, the Ego and the non-ego coincide and are seen to be merely one-sided differentiations of the Absolute Idea:

* * *

[109]

Kant showed the insufficiency of the empiricism of Hume by pointing out the universal and necessary character of the categories, while by acknowledging their purely subjective nature he annihilated the ordinary metaphysical doctrine of the Absolute: it remained for Hegel to show men that the Absolute and the Infinite were not unknowable as Kant would have it, and yet that man does not as Jacobi thought naturally by faith and intuition know the secrets of God: to him the road to true knowledge passed from the onesided identification of the understanding, to the disintegrating method of the free dialectic, till the contradictions of thought are unravelled, and the absolute knowledge of absolute Being, the Idea of pure thought is ultimately reached.

* * *

[111]

Socrates and Kant brought philosophy back to man: Aristotle and Hegel set out again to reconquer the World—

<u>Kantian Knowledge</u>

Knowledge to Kant was the thinking of our sensations[.]

Hypotheses

The historical progress of every science depends on the criticism of hypotheses—on the gradual stripping off of what is untrue or superfluous.

Thought: like Digestion

The Greeks of the 4th century were afraid that a denial of universals meant an annihilation of thought, just as in modern times Humes annihilation of cause & effect produced an intellectual panic; but Thought like digestion goes on very much the same notwithstanding any theories that may be held about it[.]

* * *

[114]

[A small (1 1/2″ × 3/4″) sketch in pencil of a man in Renaissance costume appears in the bottom left-hand corner of the page.]

* * *

[115]

The End of all things

Is there a limit to the [The remainder of the page is blank.]

* * *

[119]

Moral chemistry

Hume explained the idea of a cause by means of the law of the association of ideas: and Hartley who followed him propounded the notion of ''mental chemistry'' in which the result of two linked impressions might not put in evidence either of the components any more than water shows us the oxygen and hydrogen it contains[.]

Now it is possible that this may hold good in the moral as well as in the intellectual world: The basis of our moral Feelings may be found in the desire for self-preservation•

But this preservation of self is not the individual self but what Clifford calls the ''Tribal self'': individualism, private property, and a private conscience, as well as the nom. case of the personal pronoun, do not appear till late in all civilizations: it is the Tribal self wh. is the first mainspring of action, and canon of right and wrong: a savage is not only hurt when a man treads on his own foot, but when the foot of the tribe is trodden on.

✱ ✱ ✱

[121]

Remorse then first appears in the following form "In the name of the tribal self I do not approve what I have done" approbation similarly: now the force of categorical imperative of the modern moral sense rests on the fact that natural selection has left to survive only those tribes in whom a tribal sense is strong• and also by the strong punishments inflicted on any violation of tribal rules.

and neither in its matured developement, or in its earliest starting point of evolution, is the canon of ethics the greatest happiness of the greatest number but the preservation of the Tribe, its highest excellence• nor is goodness merely altruism, or service to others because they are others, but service done to a community by an individual who is part of that community•

✱ ✱ ✱

[123]

Lucretius rejected the idea that the origin of religion was to be found in the "solar theory" on psychological grounds, and found it instead in the phenomena of dreams• here is he corroborated by Tylor who sees in dreams the origin of animism which was undoubtedly a spiritual hypothesis tho' it may have been as Clifford suggests a biological one Too.
The Greeks presupposed a certain metaphysical idea that of attraction & repulsion to account for the molecular action of atoms: modern science tries to give a materialistic explanation by saying that carbon has four hooks, and hydrogen only one—This however seems a worse hypothesis than the first[.]

✱ ✱ ✱

[124]

Tacitus says "the images of Cassius and Brutus were conspicuous by their absence": But Dante places the murderers of Caesar with Judas Iscariot in the lowest circle: it is in Shakespeare we find the most rational estimate of them•
Catiline was hero to the young tyrannicides of the Renaissance before he became one to Mr. Be[e]sley.

✱ ✱ ✱

[125]

The use of imagination in religion is to bridge over the chasm between the God of conscience, and the God of nature, the God of the "starry heavens above" and of the "moral law[.]"

But modern science has shown us that both ethics and motion are results of molecular action: motion in one direction may be an ellipse, in the other a moral sentiment[.]

Clifford says we never can know things universally true: even the accumulated experience of 150 million years would not be sufficient to warrant us in ascribing universal necessity to any phenomena•

Now here he is at one with the idealists• but he differs from both schools in the idea that this universal necessity is a delusion, like the idea that water or the Zoetrope is continuous• when the microscope shows the granular molecules of the one, and the simple act of stopping the machine the discrete objects of the other•

* * *

[127]

From Wordsworth

To Duty

"The most ancient heavens by thee are fresh and strong[.]"

"The ideal path of right more fair than heaven's broad causeway paved with stars"

"He hath overleaped the eternal bars" (of right)

One impulse from a vernal wood
May teach you more of man
Of moral evil and of good
Than all the sages can

"We murder to dissect"

"Heaven lies about us in our infancy"

"come forth into the light of things
Let Nature be your teacher"

"Two voices are there—one is of the sea
One of the mountains—each a mighty voice
They were thy chosen music liberty"

(This could be said of the Greeks)

* * *

[129]

Attempt of the spontaneous generators to produce "hay bacillus" a definite organism from dead dust has failed—

The barren phenominiistic philosophy of Mill and others has been qualified by metempirical considerations, just as the primitive utilitarianism of Bentham has been so qualified by Mill himself as to reappear in such entirely new form that no amount of mental or moral chemistry is enough to discover the primitive germ. Berkeley annihilated the non-ego—Hume did the same for the ego—and when this was followed up by the resolving of the laws of cause and effect into a mere association of subjective ideas all honest folk thought the world was coming to an end•

* * *

[130]

[This entry appears opposite "The Bacchae of Euripides," on p. 131.]
cited by the orators as a patriot, by Aristotle as the most tragic of poets, he was to the age of Menander the model and the delight[;] more than this, Euripides witnessed to nature in the stilted rhetoric of the Roman stage, in the studied pomp of the French Court: He fed the youth of Racine and of Voltaire, fanned into a flame the genius of Alfieri, and occupied great poets as Shelley, Schiller, Browning with the task of translating him.

And we who Toil in the heated quarries of modern life may perhaps—or is it only a fancy—gain some freedom of soul from his genius who was the great humanist of Hellas, the cor cordium of antiquity[.]

* * *

[131]

The Bacchae of Euripides

when the actor substituted the head of Crassus for that of Pentheus in the tragedy of the Bacchae, the material triumph of Orientalism and the sovereignty of the Parthians began: and transformed into the Christis patiens this play formed another link between East and West—and symbolized another triumph of Orientalism the Religion of Christ.

Parmenides•

Founder of Idealism and of dialectic, of metaphysics and logic—Plato recognised him as his spiritual Father "whom he revered and honoured more than all other philosophers put together"

* * *

[132]

[This entry appears opposite "Metaphysics" on p. 133.]
So even David Hume the precursor of the modern phenomenistic empirical school of thought said "We must cultivate ^true metaphysics with some care, in order to destroy the false and adulterated"—

* * *

[133]

Greek Materialists

Plato talks about the materialists as "earth-born giants who grasp oaks and rocks and wont believe anything they cant hold in their hands", and of the idealists who "prudently defend themselves from a fastness in an invisible world". (Jowett· 4· 396)

Metaphysics

it is really not a question between ~~good~~ emploing [*sic*] metaphysics, or not, but between emploing good or bad metaphysics:

There are more metaphysics (i.e. more a priori assumption) in an unmetaphysical age than in any other· We do not consider how much metaphysics are required to place us above metaphysics—see (Jowett Plato. 4. 155)[.] We talk about facts when we are really resting on ideas:

* * *

[135]

Knowledge as Plato said consists "not in the feelings or affections (παθημασι) but in the reasoning about them (συλλογισμῳ)." The senses are not "mere holes in a wooden horse" but the organs of a presiding nature—we perceive "through them" not "by them":

Sensation seems to us simple: it is really most complex: till the mind applies its forms it is a "*συγκεχυμένον τι*"[.]

* * *

[137]

X
Necessary Truths: unwarrantable psychological facts
The experience of the race having been substituted for the experience of the individual, necessary truths are admitted to be a-priori to the individual, though a-

posteriori to the race: upon the other hand it is acknowledged now by the experiential school that no finite number of years can logically warrant the necessity and universality which we undoubtedly attribute to certain truths: "the accumulated experience of my ancestors for one hundred and fifty million years is no more competent to tell me that the angles of any triangle are equal to two right angles under all possible circumstances, than my experience of the last five minutes" (Clifford)[.]

This universality, this necessity, is an extra-logical psychological fact, resultant of a purely automatic act of the mind: it is not a logical conclusion from adequate premises—

We express our belief in logically unjustifiable language—a universal statement is really a particular statement about the nervous apparatus of thought.

* * *

[139]

admit that a universally necessary statement is valid and you have no alternative but to accept the existence of a-priori truths; a-priori to the race as well as to the individual—

Huxley says a man is lucky who has gone through life without accumulating a store of necessary beliefs which he would give a good deal to be able to disbelieve• To him necessary truth is

(1) either, a convention of language

A = A•

or (2) propositions the negation of which is

(α) inconceivable—

(β) a denial of some fact of immediate consciousness

now as regards inconceivability Mill rejected it as a mere accident of our mental apparatus, and Clifford showed that our immediate consciousness is no valid test, for our immediate consciousness testifies that water is continuous, or the motion in a wheel of life—

* * *

[140]

Herbert Spencer rests the whole of his philosophy on the test of inconceivability—"it is inconceivable that Being should pass into not-Being therefore force is always persistent:" not-being is "incapable of mental presentation", "presents no contents for consciousness."

* * *

[141]

As regards however inconceivability we often deceive ourselves—We say for instance space is either infinite or has a boundary—but both of these are inconceivable ideas—

Now the question has space a boundary is simply unmeaning, but the question has space a finite or an infinite number of Cubic miles is quite intelligible—(see Clifford 1.154.).

* * *

[144]

Science and Hellenism•

Goethe owed his sublime cheerfulness to his scientific training—For us who do not "walk delicately in sunlit air", or "garlanded with white-poplar run races in the prime of spring when the plane tree whispers to the pine" the Greek attitude can only be gained by a recognition of the scientific basis of Life—

Life came naturally to the Greeks as a very beautiful thing, we "whom the hungry generations tread down" barely can attain to the gladness that was their immediate heritage[.]

* * *

[145]

Hellenism•

In the modern attempt to rest morals on a scientific basis, as well as in Baudelaire's passionate cry "O Seigneur! donnez moi la force et le courage [/] De contempler mon coeur et mon corps sans degout" A return to old Hellenic ideal *ζήν κατα φυσιν* can be seen.

Line of Dialectic

Philosophy began with assertions about physical science and seems likely to end in the laboratory•

Heredity

Religion tells us that the father has eaten sour grapes and the children's teeth shall be set on edge—and the latest word of modern science is that the fact of our ancestors having held peculiar views on the three angles of any triangle is an inheritance from which we cannot escape—

* * *

[147]

To say• that mind is a function of matter, or that matter is the realization of mind, are propositions equally meaningless so far as they postulate a knowledge of anything more than phenomena:
Whether extension be visible thought, or thought invisible extension is a choice of words not of conceptions.
The highest functions of the nervous system are those to which the supreme hemispherical ganglia minister, organs essentially different from the sensory ganglia:

Flourens took out the hemispherical ganglia from a pigeon and the bird lost all intelligence, though a decapitated frog will by means of the spinal cord perform a series of actions which are someth[ing] more than reflex motor•

* * *

[149]

We do not enough distinguish between the different kinds of matter, and of motion•
There is no need to assume an immaterial liver behind the hepatic structure.

* * *

[150]

Plato founding his ideal state might rightly say "*το μεν ωφελιμον καλον το δε κακον αισχρον*" for utility can be the only bond for cities to rest on, but the true Greek idea comes to us in a song long before Plato's time
οτι καλον φιλον εστι, το δ' ου καλὸν ου φιλον εστι
yet notice Theocritus
"*ουχ αμιν τα καλα πρατοις καλα φαινεται ημες*"

* * *

[151]

Metaphysics•

Those who object to Metaphysics often mean no more than this that all our knowledge must be expressed in terms of the means by which knowledge is acquired, that is in terms ultimately reducible to statements about our sensations therefore strictly relative•

Metaphysics seems to me the one science which has a future• for the acquiring of the new methods of science and reasoning to which we look forward must rest always on a metaphysical basis.

* * *

[152]

[This line appears opposite the remarks on Shelley on p. 153.]
His own words about the Niobé.

* * *

[153]

The Plastic Spirit of Greek Literature

When Aristotle said the Platonic ideas were a mere doubling of the sensible world, he laid his finger on the weak point of the earlier Platonic metaphysics, but on the other hand he may be said to have touched unconsciously on the peculiar characteristic of the Greek spirit—its plastic character—it's Tendency To conceive ideas always under the form of images, and to give a sensible rendering to objects most removed from sense: this may be called the artistic spirit of Greek Literature, and may be paralleled by the attitude held by Keats towards Wordsworth who as Shelley said always seemed to "awake a sort of thought in sense" and towards Shelley who dwelt chiefly on what was vague and spiritual, and for whom the sight of a Greek piece of sculpture touched "the most removed and divine of the chords which make music in our chords"

* * *

[154]

X X X

So also Euripides makes one of his characters say
"Stand off and view my sorrow as a painter might" though Euripides perhaps of all the Greeks had the most share of the modern vague spiritualistic tendency—the tendency of Werther, and René and Faust—the morbid analyzing faculty[.]

XX

Note in connection with this how the Furies of Orestes have a tangible visible shape—and are described with a sort of Pre-Raphaelite frankness of detail, as opposed to the Furies unseen of men which Hamlet bears within his heart[.]

* * *

[155]

So in Greek Literature we find from the outset a peculiar definiteness of conception, a clearness of outline which are the plastic conditions of art, so that Phidias when asked what conception of Zeus he would carve, contented himself with repeating three lines from Homer as the answer•

We may compare Shelleys Ode on the Cloud with the celebrated Cloud Chorus in Aristophanes—the latter full of the mythopoeic and sculptural power of vivid realisation, as well as of accurate observation even to the (πλαγιαι) the side long way clouds creep down a mountain—or compare again the Hero and Leander of Musaeus with that of Marlowe[.]

In the former the motive is presented as it were from the outside—images are given to us clear in their plasticity like a statue bathed in visible sunlight—There is very little psychology—a branch of knowledge in which the Greeks were always behind hand both in the drama and in philosophy.
[The following is written at a slant in the left margin between "Leander" and "clear."] cf. also Shelley's Adonais with the Lament of Bion•

* * *

[156]

X X X

compare Aristophanes particularily [*sic*] for this peculiar sensible symbolism—the real scales in which AEschylus and Euripides are weighed, the mortar in which War and havoc bray up the states of Greece, are like *Κρατος* and *Βια* of the Prometheus, and entirely in accordance with the Greek attitude[.]

In modern times Dante and Durer, Keats and Blake are the best representatives of the Greek spirit[.]

Ç How much of this is due to the fact that Greek Literature was not meant to be read—but to be recited.

* * *

[157]

in fact the Greeks were always 'hubsch objectif" as Goethe said—and the wish expressed by one of the Euripidean characters that a mark or external sign could be put on people to show their character is in every way a symbol of their attitude.

Marlowe floods the subject with a flood of spiritual thought or passion—the details of the tower and the lamp either dont come in at all or incidentally—but to the Greeks the eye and not the ear or mind was the chosen vehicle of passion•

Connected with this is their plastic rendering of Landscape scenery—the Faun of Praxiteles with all the mystery and wantonness of the woods about him, the river nymph of Scopes, in whose tangled hair and melancholy eyes the restless sorrow of great waters fo[un]d its plastic utterance, these gave to the Greeks what the vague and misty splendour of Turner gives to us•

* * *

[158]

XX
So the plastic arts determined the direction of idyllic poetry determining the name and suggesting models of compact concentrated treatment. The Idylls are in literature what the vases are in art•

* * *

[159]

and perhaps in the personification of abstract ideas, at once the germ and the weakness of early Greek philosophy, we may discern the workings of the same spirit—

And as regards the drama when Euripides says 'look on my sorrow as an artist'' he touched the quality of the old aesthetic ideal—
Aristotle said of tragedy that it proceeds by action not by narration and a character in a Greek play is seen by his relations to others not by his soliloquies—that is[,] is viewed externality: so also Greek sculpture does not obtrude its muscular structure[.]

Between ancient and modern dramatic art there is as wide a gap in this respect as there is between the figures on the Medicean tomb—and the calm serenity of the Parthenon frieze—

* * *

[161]

Now all this may be connected with their exquisite sense of form, their dislike to *το απειρον*—to the *ερως των αδυνατων*, the sense of the infinite which is the legacy of mediaevalism to modern life, and which music the essentially modern art as sculpture was the essentially Greek art: they always sought to externalise, to realise, to remain within corporeal limits: in reading Greek mythology we are apt to forget that we are reading about beautiful natural objects, and spiritual conceptions submitted to corporeal conditions•

Modern art appeals directly to the emotions, aims at reaching the spiritual reality of things, cares more for feelings than for form—Greek art remains on the surface and translates into marble the humanised aspects of the external world[.]

* * *

[163]

Art then may be regarded as the natural expression of the spirit of man projecting itself in the forms of beauty most characteristic of its nature•

So to the Egyptians architecture—for humanity had not yet become conscious of its own beauty—to the Greeks sculpture which corresponds to the unperplexed emphatic outlines of Greek humanism—painting to the mystic depth and intricacy of of [*sic*] the middle age—and music and poetry—the sense of the infinite, and the scientific spirit of criticism to modern life—

The hierarchy of the arts like the Philosophy of History must be founded on a psychological basis[.]

* * *

[165]

The bridge between Greek and modern life may be found in the conflict of Greek tragedy—and also in the colourless abstraction of the divine forms of Greek sculpture, which seem a premonition of the fleshless refinement of the pale mediaeval painter.

* * *

[167]

Of the quality and motive of Tragedy as opposed to sculpture the Greeks were keenly conscious—

The calm Gods of the Parthenon looked down impassively on the passion of the Dionysiac Theatre•

<u>The Idyllic Tone</u>•

appears first in the Homeric hymn to Pan—then perhaps in the Bacchae and Ion of Euripides and particularly in the opening chorus of The Cyclops—the most Theocritean passage in Gk. Lit• before Theocritus[.]

And the Electra of Euripides is the first real recognition of the pathos of common life—

* * *

[168]

[The following entry appears opposite <u>Metaphysics</u> on p. 169.]

<u>Metaphysics</u>•

Like the Sphinx• those who do not try to answer her questions are annihilated—and often he who solves her riddle comes to no good end—for he dwells apart

from men in the colourless abstractions of a spiritual world, in the vacuity of an infinite which has no contents[.]
The danger of metaphysics is that men are often turning nomina into numina.

* * *

[169]

In Raffaelle and in Perugino the religious spirit smiled Through it's tears (Hegel).

Metaphysics

"We must renounce metaphysics if we would mould our lives to artistic perfection[.]

For Philosophy serves culture not by the fancied gift of absolute or transcendental knowledge but by suggesting questions which help one to detect the passion and strangeness and dramatic contrasts of Life•"

Yet surely he who sees in colour no mere delightful quality of natural things but a spirit dwelling in things is in a way a metaphysician.

* * *

[171]

Necessity

Necessity is not as of old an image without us with whom we do warfare: it is a magic web woven through and through us, like that magnetic system of modern science. more subtle than the subtlest nerves yet bearing in it the central forces of the World—

When we combat it, istis [it is?] with the weapons of its own forging, and the assertion of liberty is only the claim that from certain forms of conscious volition certain results will inevitably follow• We are indeed compassed by the high unver [never?] leaped mountains of necessity, but for him who knows his limitations this dark horizon becomes the sunlit circle of duty[.] The cross of Christ will cast no shadow when the sun of Truth is high in the heavens.

* * *

[172A—recto]

[inserted page from a smaller (7 3/8″ x 8 3/4″) notebook.]
The end of life is not action but contemplation, not doing but being: to treat life in the spirit of art is to treat it as a thing in which means and end are identified—
To witness the spectacle of life with appropriate emotions[.] To withdraw the

thoughts from the machinery of life to fix them with appropriate emotions on the great facts of human life wh[ich] machinery does not affect[.]

The end of life must be realised through the means.

✱ ✱ ✱

[172A—verso]

Those who object to metaphysics often mean no more than this that all our knowledge must be expressed in Terms of our knowledge that is in statements ultimately reducible to statements about our sensations, therefore strictly relative[.]

Spinoza said Truth is its own test, the clearness of our ideas—a clear and distinct idea proves itself[.] This is Spencer's inconceivability of contrary[.]

Spinozas Ethics give us Evolution—the continuous adjustment of internal to [e]xternal relations.
[The lower left corner of the page is torn off, the line of the tear going through the letter "x" in "external."]

✱ ✱ ✱

[173]

Progress mental and physical

The thoughts of men are widened with the process of the suns and this process has been accompanied by a progressive developement in size and complexity of the cerebral hemispheres[.]

Ideals.

Shelley in a fleet pursuit of far-off untried ideals forgot the undying beauty of things that fade and die[.]

Plastic Art

can accept no scheme of life in exchange for life itself• it cannot dream of a heaven where there is no resurrection of the body—For the painter and the sculptor there is no escape from the bondage of the earth—nay there is no desire of escape.

✳ ✳ ✳

[175]

Contemplative Life.

Rejected by Bacon because "radios nullos sive caloris sive luminis in societatem humanam diffundat[.]" He anticipated much modern speculation in his recognition of the public over the private good•

Future of Science.

The words of Bacon, true in his own day are truer still in ours—there is for us more mystery in the world than for him• "versamur adhuc plane in atriis naturae."

Experience.

Bacon complained that Experience was not yet reduced to writing[:] "neque adhuc Experientia literata facta est" But Herbert Spencer has taken this reproach away•

✳ ✳ ✳

[177]

Against Hegels Philosophy of History Bacon says (ref. to Arabs & Schoolmen) "Sunt enim non minus temporum quam regionum eremi et vastitates"
Limit to investigation of Nature "subtilitas naturae subiltilitatem [*sic*] sensus et intellectis multis partibus superat"
Scientia et potentia humana in idem coincidunt:
ipsissimae res veritas et utilitas.

✳ ✳ ✳

[179]

To Shelley the bees in the ivy bloom are scarcely heeded: they become

> Forms more real than living man
> Nurslings of Immortality.

In the full tide of the Renaissance it did seem as if in the hands of Michel Agnolo sculpture and painting were to be made again the plastic symbols in which all the utterances of humanity were to find voice: But they could only be for the South where the human body was seen and was loved—

✻ ✻ ✻

[181]

"When an opinion leads to an absurdity it is certainly false: but it is not certain that an opinion is false because it is of dangerous consequence"

now humanity has always been exactly of the opposite opinion.

Design

We have outgrown the theory of design and talk easily of the "silly maladaptations of organic nature"

(Clifford)

Motive

That we have a choice of motives is evident from the existence of vaso-motor nerves, whose function it is to get the arteries pinched•

✻ ✻ ✻

[183]

Heredity

Spencer remarks that psychical environment is as important as physical: we cannot lay the ghosts of dead creeds[.]

Κοσμος•

"A unity underlying all facts is at once the ideal and the postulate of knowledge"

Mysticism•

Philosophy passes into mysticism because it cannot answer it's own questions: the highest truth of philosophy is rational and self-conscious poetry: and the highest poetry is often irrational and unconscious philosophy•

So Virgil longed to be a physical philosopher, and Shelley ranked poetry below moral and physical science to which he aspired[.]

✻ ✻ ✻

[185]

Beauty•

"Rien n'est vrai que le beau" Beauty may be strange, quaint, terrible, she may play with pain as with pleasure, handle a horror till she leaves it a delight.

Art is one though the service of art is diverse• Beauty also may become incarnate in a myriad of divers forms but the worship of beauty is simple and absolute•

As it is the crown and prize of life—the flower which fadeth not, the joy which never disappoints—so it [is] the aim of early education•

Let a boy says Plato from his childhood find things of beauty a delight (*ευθυς παιζοι εν τοις καλοις*): and in another place he says the end of music is the love of beauty (*δεῖ γαρ τελευταντα μουσικα εις τα του καλου ερωτικα*) and these expressions come in a scheme of the noblest education—

La beauté est parfaite

La beauté peut toute chose

La beaute [*sic*] est la seule chose au monde qui n'existe pas a [*sic*] demi

* * *

[187]

Clifford accounts for the a-priori probability of Euclidean space by the fact that of the three kinds of three-dimensioned space—the hypothetical space of Euclid with no curvature—the space of Lobatchewski with constant negative curvature—and the space of modern scientists of constant positive curvature, only in the first are the three quantities specifying a point actual distances of the point from three planes. [Two parallel lines in pencil have been added in the right margin as emphasis for the closing remark about the three quantities.]

* * *

[189]

If Philosophy aims at doing good to man it comes too late in the day for that: for while religions preside over the birth of nations philosophy often follows them to their grave• it is not till the twilight comes that the owl of Athena begins it's flight—

So in Aristotle the philosophers life is the contemplative life: he expressly disavows any philanthropic aim to *Σοφια*•

"*η μεν γαρ σοφια ουδεν θεωρει εξ ων εσται ευδαιμων ανθρωπος*". (Eth. 6. 13). it is good says Aristotle for it's own sake because it is an "*αρετη* of the soul": the fact of it's existence is the reason for its existence. Bacon's scornful words are it's glory—Like a virgin consecrated to God it bears no fruit• its duty is to comprehend the world not to make it better.

It's sphere is the Universal—The universal Law of the Minds movement[.] It examines not what ought to be but what is• It is the possession of real wisdom, not the love of wisdom•

✻ ✻ ✻

[A sheet has been torn from the notebook between pp. 189 and 191.]
[191]

Universality and Necessity

Aristotle saw how difficult it was to gain from mere sensations the universality and necessity which are the factors as well as the product of experience•

Sensation gives us a perception of an *αισθητον* • a *τοδε τι*• as occupying a given space either locally or in time:
τοδε τι • *που* • *και νυν*•

Now knowledge informs us of the *αει και πανταχου*•

✻ ✻ ✻

[192]

[The entry below appears opposite p. 193.]

Plato's debt to Socrates•

The appearance of Socrates in the Platonic dialogues may be ascribed partly to Plato's reverence for Socrates as a teacher• partly from a desire to give a literary unity, a plastic singleness to the whole system• and partly to show the intimate connection of philosophy and life: from Socrates Plato got
(1) universally valid axioms of morals•
(2) method: discussion• *επακτικοι λογοι* which consisted of
    (1) negative side• Socratic irony
    (2) *η μαιευτικη τεχνη*• midwives art.
  το οριζεσθαι καθολου•
(3) virtue is knowledge•
(4) utilitarianism•
(5) the necessity of a state
(6) *τελεολογια*•

✻ ✻ ✻

[193]

Three stages of the Platonic ideas. (1) Logical (2) Ontological. (3) Poetical. R. the ideal aspect of the Platonic ideas: they are teleological because it is the *τελος*• of man• poetical because it comes from imagination[;] as intellectual it is verified by *διαλεκτική*• as beautiful it is apprehended by *Ερως*•
Ερως and *διαλεκτικη* as two sides of the same thing:

ερως• the impassioned search after truth, as well the romantic side of that friendship so necessary for philosophy• because discussion was the primitive method• From the love of the beautiful object we rise to the ideal ερως; from Charmides to the ιδεα του αγαθου•
Satisfied one's desire of immortality by providing for one's mental continuity.
In discussion "we wrestle to embrace, and embrace to wrestle"

* * *

[195]

In the Parmenides we have an anticipation of the Spinozistic omnis determinatio est negatio[:] "ideas are not absolute atoms but relative and cannot be apprehended without relation to contraries.["] In Sophist: Hegelian conception of not-being as equally essential To being[:] "the nature of a thing is only sum of it's relations to what it is not"[.]

* * *

[197]

Κοσμος•

exists in its fulness only in relation to the eternal subject God, who as eternal mind makes the natural world, and as idea of Good the moral universe•

Great Men•

The condemnation which a great man lays upon the world is to force it to explain him. (Hegel).

Plato•

"A Knight of the Holy Ghost"

Philosophy of Hegel•

A synthesis of science but in a higher sphere—in the ether of pure thought: before dividing into the Kingdom of nature, and the Kingdom of mind, there is an earlier problem—to determine the order in its native medium of thought•

The Idea•

The totality of the World—the reality because it includes everything: the ideality because all that is "immediate" has been crushed in the wine press and only the spirit remains.

* * *

[198]

[This entry appears opposite "Kant annihilated" on p. 199.]
Kant shattered himself on the rock of True Being the "Ding-an-sich" Hegel solved the problem of Metaphysics by turning it into Logic•

The absolute presents itself to us as a relative, and when we have seen the absolute relativity of relation we attain to the true absolute•

Being. Relative Being• The Notion
Universal. particular• Individual

* * *

[199]

Metaphysics

One of the good things of the Positive school has been to point out the insufficiency of those abstract entities wh. are little more than a doubling of the sensible world—such as essences, laws[,] forces, causes, when they profess to be more than a reflex of sensation[.]

rendered the protest useless•
But Hegel ~~re-created metaphysics~~ in his second part of Logic, the Theory of Essential Being, of substances, causes, forces, essences, matter in their essential relativity [.]

Kant annihilated metaphysics: but he acted on the hypothesis that man in his individual capacity is all in all.

Hegel made God the object of his philosophy—God the Original Synthetic Unity from which the non-ego and the ego have issued by differentiation, and in which they return to unity.

For the Ego and the non-ego are equally the results of a process—the difficulty of spanning the chasm is one made by men for themselves.

* * *

[201]

Mind:
Aristotle said "that the mind is nothing before it exercises itself"
Hegel remarked it was a "subject[,]" "a process" not a substance•
Aristotle rem.[arked] that an object must be criticised through it's own principles

not through general formulae. This serves us to see how worthless the "utility" measure is for all things[.]
what we think in Aristotle is the realism of sense, often turns out to be really the idealism of reason.

* * *

[203]

<u>Progress of Thought•</u>

Quality. Quantity. Measure.
To reduce quality to quantity is one step in the progress of Knowledge: to see that the proximate truth of the qualitative world (*ιδια αισθησις*). the world of sense proper, is in its simplest terms the quantitative world (*κοινα αισθητα*) the world of "one" and "many"
Quality: Being• Heracletos• Democritus
Quality. Pythagoras
Measure. Anaxagoras

* * *

[204]

[This entry appears opposite the remarks on p. 205.]

$_{X}X^{X}$ Hegelian dialectic is the natural selection produced by a struggle for existence in world of thought: so *ερως* and *νικη* were to physical speculations of Greece survival of fittest connected also with sex in plants and animals:
"The unity of truth, the consistency of the warring elements of the mind, the faith in the invisible, are all included in the Platonic *ερως*•"
Plato anticipates the evolutionist theories of Aristotle in his scheme of the gradual spiritualising of *ερως*, till from the blind sexual interests of plants it becomes the purely intellectual love of the purest abstraction: *ερως*• is the beginning of the mysticism of neo-platonism: it is like the fruitio Dei of the mediaeval saint, or Dante's love for Beatrice, or the hunger and thirst after righteousness—

* * *

[205]

Ερως•

The Platonic *ερως*• the dialectical energy of thought: "it is the child of wealth and Want• never poor, never rich: a mean between igno~~rance and knowledge:" (Symposium) it is the spirit of scepticism—the force of contradiction and the neg~~ative: the Hegelian dialectic• the inquietude poussante of Leibnitz[.]

Philosophy began in the Eleatic school of Parmenides and the idea ''Being alone is, Not-Being is not''.
The earliest systems were the most abstract and had least in them[.]
Modern science is aetiological: but the type of ancient science is geometry[.]

* * *

[206]

[This remark appears opposite ''The Greek idea'' on p. 207.]
The cognition of sensuous objects need not itself be sensuous.

* * *

[207]

Lobatchewski was to Euclid what Vesalius was to Galen or Copernicus to Ptolemy•
The Greek idea of the plastic unity of Life showed in their objection to versatility—in Plato and the Thucydidean ''*προς απαν συνετον επι παν αργον*•''—
Euripides—conservative in politics• free•thinker in thought•
The empiricism which sheltered itself in psychology and The association of ideas was demolished by Kant: but it has re-appeared again under the aegis of modern science and evolution—but this really only transports the problem back to the old ''tabula rasa'' of primeval man•

* * *

[209]

The Origin and progress of Ancient and modern philosophy•

Ancient philosophy began in cosmology—passed into logic• ethics• and physics—and ultimately with the Neoplatonists finished with theology—but mod. phil. began with a revolt against theology and ended with cosmology•

There is thus an exact reversal of the subject matter, but in method there is more similarity•

For the empirical and sceptical school which reached its ultimate realisation in David Hume may be paralleled by the Sophists: Kant like Socrates brought speculation back to man: not the individual, but man viewed with reference to the universal and abi~~ding~~ elements of his nature—man as the theoretical and practical measure of all things: Hegel like Aristotle set out again to conquer the world:

* * *

[212]

[The following remarks appear opposite " 'Every proposition" on p. 213.]
Positivists accord to metaphysics a historical value: as an advance on the theological age[.]

Their criticisms apply to the Scholastics with their objective entities and quiddities but not to Descartes—who opposed the false metaphysic that preceded him by real metaphysics•

* * *

[213]

Metaphysics to modern positivisism [*sic*] is a science "non moins sterile que puerile", for it seeks to force us to "poser des questions qui sont insolubles, et sonder l'absolu qui est insondable".

"Every proposition which is not finally reducible to a simple enunciation of a particular of a general fact cannot be said to offer any real or intelligible sense"—

It does not assert that the absolute is non-existant only that it is inaccessible: from the topmost wave of empiricism we have descried the far-off shore but we have neither rudder or compass to enable us to cross the untried waters which flow between•

* * *

[214]

Positivism may be described as catholicism without Christianity as regards it's social aspect, philosophically it is dogmatism without criticism[.]

* * *

[215A]

[The page numbered 215 is blank, and part of the bottom third is torn away; the next leaf, which was not numbered, but on which there is a short entry, separates pp. 215 and 217; we have numbered it 215A.]
To Pliny scribere facienda was as great as facere scribenda[.]

* * *

[217]

The Realistic assumptions of modern science.
That the external world exists, that there is a sequence of cause and effect—at the time that they tell us that all that we can know is the experience of mental phenomena:
The difficulty of mod• sc•

The consciousness of a persistent ego underlying all sensations and different from them•
How pass from matter which is that which has extension to mind which is that which thinks. certain material changes exterior to our organism are always accompanied by certain other material changes inside our organism—but this does not solve the problem of consciousness.

[END OF THE COMMONPLACE BOOK]

5

Oscar Wilde, Notebook Kept at Oxford

✻ ✻ ✻

[inside front cover]
Leroux de'[*sic*] l'Humanité• II. viii on cycles.

Mazarella "Storia della Critica"
'Tacite et son siècle' by Dubois Euchan.
Philosophie de l'Histoire by Holmé Wion[u]ski•

✻ ✻ ✻

[facing inside front cover]

Trojan War•
Eur. Orestes• 1640. pessimistic origin of. in Lecky II. 205. odd monkish notion of the war• Euripides also thought it was to check overpopulation. Acusilaus thought it all arose out of the designs of Venus•

✻ ✻ ✻

[1]

Survival of culture in Children's games, popular proverbs, and absurd customs•
Bride• lifting game among Australian children
origin of numbers shown by the numerous games of fingers. micare digitis &c. Scotch lads playing "tappie-tousie" and the game of "le petit bon homme vit encore" the former referring to slavery the latter to Theology—or "a little bird

told me'' or Midsummer Bonfires—or the Breton feast for the Dead—or the ordeal of the Key and Bible.

✳ ✳ ✳

[2]

in history what we are to look for are not revolutions, but Evolutions.

✳ ✳ ✳

[3]

Nature as Aristotle said is not full of inchoherent [*sic*] episodes like a bad tragedy•

═══

in the study of man we must put aside any ideas of extra-natural interference and causeless spontaneity.

motiveless will }
uncaused Spontaneity }

Philosophy of History must explain the past and predict the future phenomena of man's life in the world by ref. to general laws.

✳ ✳ ✳

[5]

Spiritualism an instance of survival becoming revival•
Prehistoric Archaeology: evidence afforded by Boucher de Perthes' excavations—the Drift valleys of the Somme that man was coeval with the Stone age• the mammoth and the wooly rhinoceros•
Lucretius v. 1281• quite accurate

✳ ✳ ✳

[6]

[These lines are a continuation from the bottom of p. 7.]
and (4) to extract from the historians account of an action how much is subjective, how much objective.

✳ ✳ ✳

[7]

To adjust the balance between freedom and necessity, between the influence of great men and that of the masses is the problem of the historian•

History began in poetry and has ended in philosophy—
Aristotles dictum about revolutions "*εκ μίκρων αλλ' ου περὶ μίκρων.*"—philosophy of history•
Problem.
(1) to ascertain whether a fact happened•
(2) the causes of the fact•
(3) its position in the evolution of humanity•

* * *

[8]

N.B. in this however Buckle is merely following the early French rationalistic historians—Montesquieu[,] Voltaire• Turgot and others[.]
[This remark appears opposite the entry on "the statical element."]

appears also in astrology which Comte says is the first systematic attempt to frame a philosophy of history—

* * *

[9]

R• change in the conception of history shown by Buckle remarking that the details of the siege of Rochelle do not properly belong to history•
v the statical element in history is to Herodotos [*sic*] 'nemesis', to Thucydides a certain 'cycle of human affairs['], to Christianity predestination, and to modern rationalistic historian those habits of custom and morals wh. are dep. on physical phenomena.
v Cousin said give me the lat. long [latitude and longitude] of a country, its rivers and mts and climate and I will deduce the race•

* * *

[10]

X if the hammer and shuttle cd. move themselves slavery would be unnecessary• (Aristotle)• and machinery having virtually fulfilled the condition the predicted result has followed•

* * *

[11]

Even animal monstrosities are now known not to be capricious but essentially natural• a new science is thus produced[,] that of Teratology• cf. also the now ascertained laws of the apparent irregularities of crystals• and Pinels book on the Laws of insanity•

it is impossible to lay down a railway without creating an intellectual influence—it is probable that Watt and Stephenson will eventually modify the opinions of mankind as much as Luther or Voltaire—

* * *

[12]

Lecky is the proper antithesis to Buckle—he lays more stress on the emotions, Buckle on the intellect[.]

* * *

[13]

Humanitarian sympathy constantly changing and enlarging—code of honour different in each age• so Michelet says of the psychological historian that he can assign to a specific period a moral fact or mode of thought with as much certainty as an art critic can fix the date of a particular capital—(Lecky 304). nay with more, for the forms of art may be reproduced, a recurrence of moral conceptions is impossible—

* * *

[15]

change in public opinion caused sometimes by a single book or controversy as in the case of Harvey and Copernicus—more usually by what is called the spirit of the age•
R. How Aristotle was introduced by the heretics and always connected with heresy•

* * *

[16]

the chief moral agent of antiquity was the sense of human dignity, the sense of sin took its place in modern mediaevalism—the first produces the qualities of a patriot, the second those of a saint[.]

* * *

[17]

What is morality but the perfect adjustment of the human organism to the actual conditions of Life, and to the observance of those inexorable Laws to break which is death•

in the ancient world patriotism was the presiding moral principle and not religion: Leonidas Regulus and Harmodios are the pagan parallels to Xtian martyrs•

Patriotism is the keynote to the rise of the Roman Empire• their earliest sculpture was not of their Gods but of their heroes—so also their Literature• Livy is the patriot historian.

* * *

[19]

History of the art of war•
in early civilizations military achievements chief source of dignity: but the immense superiority of cavalry to foot soldiers ensured the success of the upper classes:
the rise of infantry goes along with the rise of democracy. In England the archers always first place•
First blow to cavalry• invention of gunpowder, and of the bayonet•

* * *

[21]

The inductive method as applied to History is a modern innovation—coeval with the introduction of the same method into Science:
Philosophy of History is nothing separate from the facts of Hist• but contained in them•
History may be looked on either as a work of art—whose *τέλος* is in the maker—or as an organism, evolving its own *τέλος*•

if we take the first view we are sure to fall into the pitfall of an a-priori conclusion—the latter is the only scientific one[.]

* * *

[22]

{ Use of the poetic faculty in science•?

{ Use of the scientific knowledge in poetry?

see on <u>Shelleys</u> passionate enthusiasm for science Symonds 14: His delight when he found that the four elements were not final• his mysterious awe at the orrery• also Leonardo di [*sic*] Vinci• astrology and alchemy mark the time before poetry and science had been differentiated• so also <u>Goethe</u> who is claimed as the first Darwinian and who was a devoted pupil of science[.]

* * *

[23]

Science and poetry:
symbolised in the old fight between philosophy and poetry, between reason and imagination of which Plato talked: which was partly compromised by the division of the poets into two classes[:] the 'inspired' and the 'versatile'—in a narrower sense we may note the Euripidean love for science as opposed to the old AEschylean terror of the awful powers of nature—note also the artists love of definite conceptions in the gt [great] Catholic epic of Danté [*sic*] and in a lesser degree in the Puritan epic of Milton• note. (1) idea of vastness. (2) of law (3) of ensemble. (4) that the ultimate of known ultimates is force. (5) of [continued on p. 25]

* * *

[25]

heredity as in the Spanish Gypsy. (6) of Evolution and progress—
N. how science and democracy join hands:

* * *

[26]

In criticising Machiavelli the question to be asked is are his principles true, not 'are they moral': He employed the legitimate method of abstraction and isolated his problem from the disturbing effects of morality as a Pol. Econ does his.

For an Essay.
Social effects of a deductive as opposed to an inductive philosophy?
Rem• however how Hume and Adam Smith were deductive• not inductive:

* * *

[27]

Aristotle knew nothing of crucial instances or the theory of averages.
The rise of inductive philosophies, is democratic and encourages the middle trading classes.
a Deductive system of philosophy is always separated by a wide gap from the people and so has very little use for them: Buckle compares Germany, and Scotland and ancient Greece: (I doubt the latter very much•)
induction treats of facts first, and ideas afterwards: Deduction vice versa: Liebniz [*sic*] however said
"The attaching too high a value to mere facts is often a sign of a want of ideas"

* * *

[28]

nothing can dispense with criticism so much as a moral aim•

---

nothing is easier than to accumulate facts, nothing is so hard as to use them•

---

spirit of exclusive
the same artistic ^ attention to form which made Euripides often prefer, like ~~Algernon~~ Swinburne, music to meanings, and melody to morality, which gave to the late Greek statues that refined effeminacy, and over-strained gracefulness of attitude, was felt in the sphere of History•

---

* * *

[29]

For a scientific conception of History the first requisite is the doctrine of uniform sequence• in other words that certain events having happened, certain other events corresponding to them will happen: that the past is the key of the future:

In the character of a nation inconsistency is impossible: in the moral world as in the physical nothing is anomalous: yet while we have now a natural science in the sphere of history the theological doctrine of irregularity still holds place:

---

* * *

[30]

Plato underrated induction, Bacon undervalued deduction, the two men who can use either method with equal ease are Aristotle and Newton:
Hunter and Cullen two gt• [great] Scotchmen are the earliest pathologists: their object was to show that there is a science of the abnormal: that in the reign of Law rebellion is unlike: Historians in treating history have been pathological rather than physiological:

* * *

[31]

"We have only five senses, but science has added largely to what nature has given us in the microscope, telescope, machines for the weighing and measures of infinitesimal atoms:"

The microscope has shown us that the division of bodies into animate and inanimate, or organic and inorganic, is quite arbitrary•

Two methods.
(1) physiological: wh. investigates bodies in their normal or healthy condition[;]
(2) pathological: wh. invest. bodies in their abnormal and unhealthy condition•

* * *

[33]

Adam Smith

Theory of Moral Sentiments, he investigates the sympathetic side of our nature•
Wealth of Nations: the selfish side:
the complexity of the subject precluded the employment of the inductive method, so he employs the method of abstraction, or imaginary separation of inseparable facts, and proceeds deductively:
Geometry the gt• [great] instance of the abstract method:

* * *

[35]

Asceticism opposed by both Greeks and moderns: by the first owing to a sense of the dignity of man, the necessity of harmonious developement, and the love of physical beauty. by the moderns in spirit of utility: opposed to commercial interests•

Ancient colonies due to love of migration, excess of population, desire of conquest: the purely industrial colony was a product of the Portugese [*sic*] empire and Vasco di [*sic*] Gama•

* * *

[36]

[The entries on this page appear opposite the remarks on Herbert Spencer on p. 37.]
common sense: external world•
science: persistence of force
religion: indefinite consciousness of existence transcending relations.

Matter[,] motion and force are merely symbols of the unknown cause or reality which is coextensive with all the phenomena of the world—Progress is as much to the unknown and unknowable as to the known• ultimately all accountable and natural facts are unaccountable and unnatural:

* * *

[37]

Herbert Spencer•

Three primary truths: the indestructability of matter• the continuity of motion• and the persistence of Force: The two first are derivative, the last ultimate: However: the interpretation of all phenomena in terms of matter[,] motion, and force, is nothing more than the reduction of our complex symbols of thought to simpler symbols: and when the equation has been brought to it's lowest terms the symbol is a symbol still.

* * *

[38]

Mykenae•

Thucydides probably never visited it, though he was in Arkadeia: in the tragedians it is uniformly confused with Argos: Strabo says there are no traces of it at all• our description comes from Pausanias.

Sicily

Where did Thucydides get the accurate details about the dates of the Sicilian colonies? was there Accurate Greek Hist. for that date? if so why not for rest of Greece.

* * *

[39]

Rem. what every great man has prided himself on is, not the results he attains to, but the method he follows[:]
and it is the method which is really important: cf. Bacon and his Novum organum• Arist. *οργανον* &c:

So in judging of Thucydides and Aristotles treatment of history it is their method which is valueable [*sic*].

(______).

* * *

[40]

[This entry appears opposite the remarks on Xenophon on p. 41.]
rem• a wonderful sneer at the cosmopolitan civilization of democratic Athens. all other Gk. cities, he says, have their particular dress, habit of life, and language• but the Athenians are made up of Greeks and barbarians[.]

C̣ what pure blood was there in Greece by their own showing.

✻ ✻ ✻

[41]

Rem. Xenophon's tract de Rep. Atheniensi probably written about 430 and consequently not by him• (see Cobet and Kirchoff.) it is consequently the
(1.) earliest piece of Attic prose• it is in a sort of dialogue[;]
(2) and most awkwardly written:
(3) it is the earliest of the gross oligarchical attacks on the Athenian democracy.

✻ ✻ ✻

[42]

Note [The word is written at a slant in the upper left corner of the page.]
the poetic imagination furnishes the principles so is connected with the deductive method•

In early civilizations science is found intimately blended with poetry. cf. the cosmologies: and the religious worship of the universality of number• in the Pythagoreans•

And in modern science, the fourth dimension of space, infinity, eternity, &c are poetical conceptions•

So Ruskin shows the inherent truth in ancient symbolism: as of the all pervading power of atmosphere symbolized by Arg[us] among the [continued on p. 44]

✻ ✻ ✻

[43]

Use of the poetic faculty in science?

Rem• how the early Greeks had mystic anticipations of nearly all great modern scientific truths: the problem really is what place has imagination and the emotions in science: and primarily rem. that man must use all his faculties in the search for truth: in this age we are so inductive that our facts are outstripping our knowledge—there is so much observation, experiment, analysis—so few wide conceptions:

we want more ideas and less facts: the magnificent generalizations of Newton and Harvey cd. never have completed [*sic*] in this mod. age where eyes are turned to earth and particulars[.]

✻ ✻ ✻

[44]

[continued from p. 42] oldest Gods, the modern truth that the blue of the air is due to reflection as symbolized in the blue eyes of Athena answering to the blue aegis on her breast• So also on Homers metaphor "brave as a fly"
"The attaching too high a value to facts is often a sign of a want of ideas•"

✳ ✳ ✳

[45]

Buckle points out how the 17th century was the age of Newton and Shakespeare—

Poets present to men of science splendid ideals to reach to: Shelley if he had lived might have given to modern thought this elevating character•

The transcendental movement has its influence on science as well as on poetry and religion•

Two physical philosophers who deliberately cultivated the imagination are Black and Leslie: on Heat. the physical manifestation of the God of Poetry[,] Apollo[.]

✳ ✳ ✳

[47]

Distinction between organic and inorganic matter is founded on the empirically ascertained facts that in the first there is greater complexity of molecular action, cellular growth and structure, and a predominance of carbon as opposed to silicon: These however may be only steps to a higher synthesis or developements from inherent potentialities in what we call inorganic matter[.]

✳ ✳ ✳

[49]

Miracles•

Hume's argument from probability essentially unsound• It may be improbable that water ever turned into wine• but it is not impossible or inconceivable•

a Law of nature is merely often verified experience•

Upon the other hand the more a statement of facts conflicts with ordinary registered experience the more complete ought the Evidence to be• it is merely a question of Evidence•

To define a miracle as a violation of the Laws of Nature is absurd• Nature is all which is: it is the series of phenomena of which the alleged miracle is one.

✳ ✳ ✳

[51]

Comparative anatomy shows us that, physically, man is but the last term of a long series which leads from the highest mammal to the almost formless speck of living protoplasm which lies on the shallow boundary between animal and vegetable life• so does comparative psychology or the anatomy of the mind•

✵ ✵ ✵

[52]

[These three remarks appear opposite the paragraphs on p. 53.]
All governments rest on Opinion•
Social and bodily organism
Kants view of Knowledge

✵ ✵ ✵

[53]

› Force is always on the side of the Governed• all governments rest ultimately on ~~experience~~ opinion •

› the social organism resembles the bodily organism, but not the individual, but the generic type•

the pearly Nautilus has lasted from the Silurian epoch to now• so may social institutions.

---

all knowledged [*sic*] is founded on experience of sensations but is not all derived from those experiences[,] for the impressions of relation have a potential or a priori existence in us and by their addition to sense experience constitute knowledge•

✵ ✵ ✵

[55]

object of philosophy according to Kant is what can I know? what ought I to do? what can I hope for? the last two resolve themselves into the first: and this philosophy must rest on science and psychology. While psychology rests on the physiology of the nervous system• the popular notion of mind is that it is a metaphysical entity seated in the head like a telegraph operator—modern science contends it is a function of the brain•

✵ ✵ ✵

[56]

is all human sorrow as meaningless as sea sickness•?
is the voice of one crying in the wilderness merely the result of the molecular action of locusts and wild honey?

Sanctity can be discovered by the microscope, and faith analyzed in the spectroscope, but this is nothing more than an expansion of the ordinary fact that a man's face is red if he is angry•

* * *

[57]

Matter and Mind

matter is what possesses extension, while spirit is that which thinks, so the spirit must be an indivisible entity. All changes of matter being modes of motion[,] how can we account for the action and interaction of a moving extended material body and an indivisible entity:
the Explanations given.
(1) The Occasionalists: God continually interfering
(2) the Leibnitz preestablished harmony: clock metaphor•

* * *

[59]

(3) the Berkeleyan hypothesis of abolishing the substance of matter altogether: (The same arguments however wd justify the abolition of soul.[)]
(4) Spinoza asserted there was only one substance whose two chief attributes are thought and extension.
The latter pantheistic hypothesis can hardly be maintained: the rain drops which by the force of molecular action run into frost ferns on the window pains [*sic*] can hardly be said to have consciousness: so modern scientists take refuge in agnosticism• or carry out Berkeleys principles to their logical result and assert that mental phenomena

* * *

[61]

no less than corporeal phenomena are merely phenomena and that in either case to predicate substance is a [*sic*] unwarranted assumption.

In the latter case however we can speculate whether mental phenomena are coaeval with corporeal: or whether they can exist prior and subsequent to the body• whether there be not dispersed a spiritual substance which nature uses as she employs matter• and which loses its memory on each incarnation• On the whole it is interesting to note that while modern scientists object to any conception of the soul as a spiritual entity which can be thrown out of the window, or a ghostly [continued on p. 63]

* * *

[62]

[This entry appears opposite the remarks on ''motion'' on p. 63.]
Δῖνος is the new God ''*Δι' ϵξϵληλακως*'' and the God Dagon may be taken as the type of modern evolution

* * *

[63]

[continued from p. 61] telegraph clerk sitting inside the brain, they are not prepared to assert that thought is coextensive with molecular action, or that there is only one substance, whose attributes are thought and extension. They urge however that as evolution has put an end to any necessity for a creator, so motion being discovered to be the primary and sole quality of matter there is no need of imagining a mover or soul:

* * *

[64]

[These notations appear opposite the appropriate entries on p. 65.]
Xenophon•
Facts• their importance and their complexity•

* * *

[65]

Gibbon admired Xenophon for his style, his sentiments and the practical lessons which may be drawn from him•
D'Alembert proposed that at the end of every century a selection of facts should be made and the rest burned: Gibbon remarks ''Let us preserve them all most carefully: a Montesquieu will detect in the most insignificant facts relations which the vulgar overlook•[''] This question really deals with a mut [most?] important one, the growing necessity for a deductive method for historians owing to the increasing complexity and number of facts•

* * *

[67]

Indian History

The Hindus had no taste for history or chronology• their acute analytical and logical mind was devoted to grammar, criticism and philosophy•

Their imagination ran wild in history• our only clue is the identification of the Indian Chandragupta with the Greek Sandracottus•

Chinese History

produced under the protective influence of Government interference and surveillance and so is without the very first essential of history[,] the questioning spirit of freedom and revolt against authority[.]

* * *

[69]

Political Messiahship

How far was this doctrine held in antiquity
Plato's philosopher King who was to be the panacea for all social evils is clearly a Political Messiah: and Thucydides had he not been averse to supernatural causes wd. have regarded Perikles as one. In Roman Hist: Polybius preaches the doctrine of the Roman state, under the empire it was the commonplace of the ordinary adulation of each new monarch•

* * *

[71]

''Dull insensate matter''
''cold dry rationalism''

These expressions are, or ought to be, relegated to the already overstocked collection of ignorant theological abuse: Motion is the essence of matter, and progress the result of rationalism.

* * *

[73]

It is rem. that with the rise of the philosophy of History in Greece was a feeling of cosmopolitanism chiefly among the Cynics• the same thing occured [*sic*] in Germany at the time of the German Illumination in such men as Goethe, Frederick, Lessing, Fichte• and was a strong feature of the Italian Renaissance•

State of nature• means either

(1) the lowest possible state in which a man can exist continuing to be human• (Hobbes)

(2) his highest developement

* * *

[75]

Polybius prides himself very much on his military knowledge: so does Gibbon who says that the captain of the Hampshire grenadiers has not been useless to the historian of the Roman Empire•

Why histories are so soon superannuated• partly because the attitude of mind, the point of view continually shifts: we are always adding ourselves new facts to the social organism which essentially alter our stand point• the Reformation[,] the French Revolution entirely change the light by which we look on history• this is the chief reason, there are of course minor ones such as the discovery of

* * *

[77]

new materials such as M•S•S•[,] coins[,] inscriptions and the like, the employment of a new method such as the comparative method, or the closer study of old authorities•

Now the test of a great historian is how far can his work be regarded as an *αγωνισμα ες αει*• every one of course represents the spirit of his age, but there is an eternal aspect of the Spirit of every age which may be caught• To recreate the past from the mutilated fragments of the present is the task of the Historian: and note that the more remote an age is the more important are the social and general aspects of that age• After all individuals [continued on p. 79]

* * *

[78]

x[X] so in Greek tragedy the choruses express attitudes of thought, and turns of mind which it is difficult for us to sympathise with: while the actors deal more with problems that concern ourselves•

* * *

[79]

[continued from p. 77] are always more or less The same• it is the backgrounds[,] the social surroundings[,] which differ:

the fulness and harmony of the facts of a great epoch must be viewed from a certain height and distance to be completely apprehended• as the formation of a mountain, its outline, relation to other chains, and general lines of configuration can be studied better from a far distance than by climbing it: Zeus could form a better opinion of the battle round Troy from Ida than the combatants themselves[.]

* * *

[80]

[These remarks appear opposite the entries on p. 81.]
Importance of Facts:
Polybius remarks that a good historian will select a fact as a workman does a sample•

* * *

[81]

as the anatomist can from a single bone create the whole structure of those marvellous animals now extinct, or as the Botanist from a single plant can infer the general character of the Flora of a district, so to a mind truly endowed with the historic sense the most insignificant fact is often [of] the highest importance•

Facts are the oil by which the historic muse feeds her lamp: but the light is given by ideas:

Facts are the labyrinth: ideas the guiding thread•

* * *

[83]

the ordinary appeal to experience is in reality a recognition of the ultimate basis of a philosophy of History• namely that from the same causes the same results will happen and though no fact is really the same as another[,] just as no leaf has its actual counterpart[,] still there may be the closest similarity[.]

* * *

[85]

<u>Didactic Element</u> in thought

R. the modern objection to the didactic element in art and poetry and science• also in History: this is a natural reaction from the continual interference of ethics into the sphere of the pure intellect under the aegis of theology•

History has splendid lessons doubtless: but they must not be inforced [*sic*] from without• the ancients regarded rather that width of mind which is experience and culture was to be the aim• They never perverted History to preach a creed—

* * *

[86]

Columbus and Copernicus revolutionised the aspect of the mind of humanity• The one sailed to the new world with the words of Aristotle[,] Strabo and Ptolemy fresh in his memory• a fragment of Philolaus set the other on that train of thought which has degraded this planet.

* * *

[87]

the study of Greek implies the birth of criticism, comparison and research• except the blind forces of nature nothing moves in this world which is not Greek in its origin.
As regards the coincidence of art and lax morals Symonds says the historian notes synchronisms whereof he is not bound to prove the interdependance [*sic*] and between which he may feel there is no causal link•

* * *

[89]

R. the essentially plastic character of the Greek thinkers whose works are penetrated through and through with a specific quality• a singleness[,] a concentration which we contrast with the comprehensive width of the modern mind, which has gained in criticism what it has lost in creative power•

Now this plastic unity of type and motive is to be found in Thucydides and in Polybius as opposed to the Asiatic versatility of Herodotus•

* * *

[91]

Michel Agnolo [*sic*] alone could express in sculpture the passionate humanity and mysticism which more readily is expressed by painting and poetry•

* * *

[93]

Facts: the significance of a single Fact
the most remarkable incident in modern History says Teufelsdröckh is not the Diet of Worms, still less the battle of Austerlitz[,] Peterloo, Waterloo or any other battle: but an incident passed carelessly over by most Historians and treated with some degree of ridicule by others: namely George Fox making ^to^ himself a suit of leather.

* * *

[95]

Archibald Alison wrote ^his thirty volumes to prove that Providence was on the side of the Tories. Herodotus had a theological bias• Thucydides wrote to expose the vices of democracy• Tacitus to expose Imperialism•

✷ ✷ ✷

[97]

Science and History

Not a single branch of enquiry can be pursued successfully alone: there is a continual enlarging of boundaries, and an absorption of the results as well as the method of other sciences.
Note also• that when dealing with a past condition of things we must explain everything by the instrumentality of causes which are now in operation• the causes of past phenomena are to be sought for in the causes of phenomena present before us• This is certain a priori of our own planet• while the Keplerian gravitation law, and the discoveries of the spectrum analysis tend to show that in the most distant stars the laws and combinations of elements are the same• so it is with History•

✷ ✷ ✷

[99]

nec deus intersit is a rule of history as well as of art•

There is undoubtedly an endless variety of manifestations among the phenomena which form the subject matter of science and history• but the ultimate constituents are the same in both cases• the chemist and the Lawgiver may produce new combinations• the Botanist may change the character of the vegetation of a whole region by the artificial introduction of a foreign plant, just as the Gt• [Great] King could transplant whole peoples at will• but ultimately the essential characteristics of the phenomena, human nature in one case, external nature in the other, are the same now as they were when the ancestors of the [continued on p. 100]

✷ ✷ ✷

[100]

Greeks were Aryan shepherds in the table lands of Tibet, and the splendid Flora of Asia and America had not been introduced into Europe—we see then

(1) the phenomena are the same now as they were two thousand years ago

(2) the causes of the various phenomenal changes have not varied• no force has been lost, or added to the dynamic laws.

✷ ✷ ✷

[101]

Rise of History•

Poets are the original historians as well as the original men of science•
1st• the purely imaginative myth
2• the semimythic, where real persons are seen through the halo of poetry•
3• the chronicler• who translates poetry into prose•
This (Froudes genealogy) overlooks the connection that History has with state registers and religious lists: In Livy both can be discerned•

* * *

[103]

Gibbon

All religions are to the vulgar equally true• to the philosopher equally false• to the statesman equally useful•
The Sibyl of History has buried [burned?] many of her volumes.
most historians cannot look the sun of truth straight in the face• they form their opinions of it from the reflection of its orb in the troubled waters of their own imagination•

* * *

[105]

Physical Science

The influence of physical science is rather the attracting influence of a new analocy [*sic*] than a practical disproof of any particular belief• for most of the beliefs which have faded away into mist before the rising sun of truth are really of a nature which is as little capable of disproof as it is of proof•

But the force of analogy[,] the desire to bring all ones thoughts into harmony, and mutual correspondance, have led men to infer that the reign of Law which is the first message of physical science, is also to be extended to those phenomena which seem the most remote from Law•

* * *

[105A]

Titus Livius

[107]

The Philosophy of History under Plato[,] Aristotle and Polybius•

With Plato it was a rational a priori system founded on psychology• this system Aristotle criticises partly on general grounds but chiefly by inductive negative instances• In the hands of Polybius it became mere empiricism—success is made The Test of rationality in a somewhat Hegelian way•
One of the best points urged by Aristotle is that Socrates does not take account of the infinite variety and complexity both of the causes of revolutions as well as of the kinds of states••

* * *

[108]

Herodotus' philosophy of history was the universality of the moral Law and punishment in this world• this is a good deal wider Than any dogma such as Bossuet's•
Polybius had really a conception of a destiny of humanity in the shape of a Roman Empire and from him we may date in litterature [*sic*] that cult of the city of Rome which we find in Livy in Virgil in Dante and Mazzini•
[The following is written in pencil at a slant at the bottom of the page.]
opening of Cicero de Legibus p. 1163

* * *

[109]

The philosophy of history with Plato and Aristotle was merely the ascertaining of the laws of the phenomena, of the effects which follow certain causes: there is really no explanation of the motive of history as a whole, of the rational design which unifies the opposing elements• and so their explanation is an essentially scientific one, and is not mixed up with theological views: They ascertained and formulated the laws of the complex phenomena of political changes in Greece but of the motive, destiny, or meaning of Humanity as a whole they sought to give no explanation• and indeed any such explanation belongs rather to the seer and prophet and is to be sought for among the mystic visions of the Semites and the Persians[.]

Renan• V. de J. 49•

* * *

[111]

Montesquieu• my edition of.
193• That no really despotic power exists: the Persian King may make a father kill his son but cannot make his subjects drink wine•
121• comme la grandeur de la republique fut fatale au gouvernment republicain [*sic*], le grandeur de l'empire le fut a la vie des empereurs.
[The following two entries were made in pencil.]
19• on the free population of Athens and Rome
152 most important• we say a state is ruined by a battle• it is not so• a state is ruined by the general conditions which make it's safety depend on a battle[.]

* * *

[113]

For a metaphor• There are stars which send a blue and red light:

Herodotus: test of history psychological•
(1) no woman carried off agst• [against] her will 1.
(2) If Helen had been really in Troy Priam wd. not have been so *Φρενοβλαβης* as not to give her up. II. 118

II. 134•

* * *

[115]

Renan on the details of the Life of Christ• Ces details ne sont pas vrais a la lettre• mais ils sont vrais d'une verité superieure [*sic*]: ils sont plus vrais que la nue verite si ce sens q'ils sont la verite rendue expressive et parlante, eleveé [*sic*] a la hauteur d'une idee.
So Aristotle says that poetry is truer than history, and this is the attitude in which one ought to criticise the speeches of Herod• & Thuyd. [*sic*]

* * *

[Eight sheets following this entry have been ripped from the notebook; the remaining sheets are blank.]

[END OF THE NOTEBOOK KEPT AT OXFORD]

6

Notes to Commonplace Book

✳ ✳ ✳

[1]

value of Racine and his school Wilde has inscribed "his" over "its."

as S. Beauve . . . important to us Probable source: Charles A. Sainte-Beauve, "Lamartine" (1830), *Premiers Lundis, Début des Portraits Littéraires,* vol. 1 of *Oeuvres,* texte présenté et annoté par Maxime Leroy (Paris: Librairie Gallimard, 1956), 327. Wilde may have in mind the sentiment in the paragraph beginning, "M. de Lamartine, le seul dont nous ayons à occuper . . ."

✳ ✳ ✳

[4]

Prevision• in social . . . are only approximate Confirmed source, Herbert Spencer, *The Study of Sociology* (New York, 1873; rpt. New York: Appleton, 1884), pp. 39–40.

✳ ✳ ✳

[5]

Is it possible . . . theory of Life Probable source: Herbert Spencer, *The Study of Sociology,* summary of chapter II, "Is There a Social Science," pp. 25–47. In this chapter Spencer explains and answers each of the three objections, but the only correspondence of phrasing is "great-man-theory," p. 32 and passim.

Even however . . . there is science possible source: Spencer, *The Study of Sociology,* pp. 44–47. There is no direct correspondence of phrase, but Spencer argues that "there can be prevision of social phenomena, and therefore Social Science" (p. 46).

Every aggregate . . . unit of society Confirmed source: Spencer, *The Study of Sociology,* p. 386.

so to understand . . . understand Confirmed source: Spencer, *The Study of Sociology,* p. 390.

* * *

[7]

human nature• there . . . Laws of Life (Biology) Confirmed source: Spencer, *The Study of Sociology,* p. 390.

The science of society . . . individual and the social organism Possible sources: Spencer, *The Study of Sociology,* p. 330; idem, "The Social Organism," *Illustrations of Universal Progress: A Series of Discussions* (New York: Appleton, 1864), pp. 388–89, 391–92.

. . . Plato saw the real analogy The analogy of comparing individual to state appears throughout *The Republic.* Socrates proposes it as a way of seeing close up and at a distance (§368); he next suggests that the justice of one man and of a state might be compared (§368); and finally proposes that a true state is like a healthy individual (§372). See *The Dialogues of Plato,* trans. Benjamin Jowett, 2nd. edn., 5 vols. (Oxford: Clarendon Press, 1875), III, 238–39; 244.

For what characterises . . . and of structure Confirmed source: Spencer, *The Study of Sociology,* pp. 330–31.

The "physiological division . . . the law of the social organism Confirmed source: Spencer, *The Study of Sociology,* p. 335. See also Spencer's "The Social Organism," p. 404.

* * *

[9]

So in the history . . . Labour in the social organism Probable source: Spencer, *The Study of Sociology,* pp. 331–35, esp. p. 331. This entry summarizes Spencer's remarks on those pages.

* * *

[13]

The Sciences . . . habits of mind The remarks under this heading are based on Spencer, *The Study of Sociology,* Chap. XIII, "Discipline," pp. 314–26, passim.

From the . . . of relation Confirmed source: Spencer, *The Study of Sociology,* p. 316.

the absoluteness . . . contingent matter Confirmed source: Spencer, *The Study of Sociology,* p. 318. Spencer's phrasing is "mathematicians are bad reasoners on contingent matters."

From the . . . too analytical Confirmed source: Spencer, *The Study of Sociology,* p. 320.

From the concrete . . . fructification of causes Confirmed source: Spencer, *The Study of Sociology,* p. 324.

Abstract sciences . . . synthetical totality Confirmed source: Spencer, *The Study of Sociology,* p. 322.

* * *

[14]

Progress. . . . for self-preservation Possible source: Spencer, *The Study of Sociology,* pp. 238–39. Wilde may be elaborating on Spencer's remark: "Is it not a truism that without divergence from that which exists, whether it be in politics, religion, manners, or anything else, there can be no progress?" (p. 238).

* * *

[15]

In Sociology . . . historic medium Confirmed source: Spencer, *The Study of Sociology,* pp. 10–11. Wilde summarizes Spencer's remarks.

Matter often . . . boiling hot Confirmed source: Spencer, *The Study of Sociology,* p. 12.

the fact that . . . decrease crime Confirmed source: Spencer, *The Study of Sociology,* p. 13. "Again, is it not clear that the repression of crime will be effectual in proportion as the punishment is severe?" Spencer asks rhetorically; he then shows that the amelioration of laws decreases crime.

Light falling . . . thoughts• feelings• actions Confirmed source: Spencer, *The Study of Sociology,* pp. 402–3.

In the bodily . . . one of compromise Confirmed source: Spencer, *The Study of Sociology,* pp. 396–97, Wilde summarizes Spencer's position about the "policy of compromise."

* * *

[16]

"sarcode" The first use in English, as recorded by the OED, is 1853; Thomas Rymer Jones's *A General Outline of the Animal Kingdom,* 4th edn. (1871), is recorded in the OED as the source for the remark that "the protoplasm of Botanists and the . . . sarcode of Zoologists, if not identical, are at all events . . . analogous formations."

* * *

[17]

"Combien de ceux . . . une belle deception" "How many of those who deny immortality would deserve a beautiful disappointment."

* * *

[18]

Bathybius Haeckelii Between 1868 and 1876 this marine chemical compound was thought to be a possible source of life through abiogenesis. See Philip F. Rehbock, "Huxley, Haeckel, and the Oceanographers: The Case of Bathybius Haeckelii," *Isis*, 66 (1975), 504–33.

* * *

[21]

Criticism at first . . . formulated system Cf. the "Rise," CW, 1106.

Homeridae in Chios The Homeridae were a clan of rhapsodists, originally claiming descent from Homer, who flourished at Chios, supposedly Homer's birthplace. They were devoted to reciting Homer's works and telling stories about his life, and were regarded by other rhapsodists as authorities and arbiters.

* * *

[23]

Epicharmos Epicharmos (c. 540–450 B.C.), the comic poet of Megara, Sicily, who wrote an early form of Attic comedy in Dorian dialect; he is sometimes called the inventor of comedy (his first play was produced c. 500 B.C.).

Sophron Sophron of Syracuse (fl. c. 440 B.C.), a writer of mimes and prose dialogues in the Dorian dialect which were highly thought of by Plato.

Pherecrates Pherecrates, an Athenian comic poet, won his first victory in 437 B.C. He criticized the tragic poet Melanthius, a minor writer of the fifth century B.C. reputed to have been effeminate, gluttonous, and malcontent.

Cinesias Cinesias (c. 450–c. 390 B.C.), a dithyrambic poet of Athens, was ridiculed by Aristophanes and other comic writers.

Timotheus Timotheus of Miletus (c. 450–c. 360 B.C.), reputed to have made innovations in the theater and to have increased the number of strings on the cithara to eleven.

* * *

[25]

Tò Κάλον the beautiful. Wilde incorrectly accents the first syllyable of the second word; it should be *Καλόν*.

* * *

[29]

Aristarchus Aristarchus of Samothrace (c. 217–145 B.C.), critic, grammarian, and head of the Alexandrian Library (c. 180–145 B.C.), produced a critical edition of Homer which is still textually valuable, as well as critical editions of other classical writers.

"l'homme et le milieu" "man and his surroundings"

connexion between the . . . climate and situation Cf. the "Rise," CW, 1127.

* * *

[31]

Strabo Strabo of Amaseia, Pontus (c. 64 B.C.–c. 21 A.D.), seventeen of whose books of *Geography* survive; his literary criticism of Homer and other poets appears in the first volume.

Dionysius of Halicarnassus Dionysius of Halicarnassus (fl. c. 25 B.C.) noted for his history of Rome *(Archaeologia* I.3 contains the observation that the style is the man), his rhetoric *(On the Arrangement of Words),* and his critical works on orators and writers.

Plutarch Mestrius Plutarchus of Chaeronea (46–120 A.D.); his surviving works are *The Parallel Lives* and *The Moralia.*

Dio Chrysostom . . . "Olympiacus" Dio Cocceianus, later called Chrysostomus (c. 40–c. 112 A.D.), was the author of many orations on political, philosophical, and literary subjects. In the "Olympic" oration he presents Phidias describing the principles which guided his sculpting of the Olympian Zeus.

Lucian Lucianus Samosatensis, or Lucian of Samosata (c. 120–c. 200 A.D.), wrote declamations, essays, biographies, romances, and dialogues. His "On the Dance" (also translated "Of Pantomime") combines philosophical dialogue with techniques of occasional oratory.

* * *

[33]

Philostratos Flavius Philostratos the Elder (c. 170–c. 247 A.D.) discussed the imagination in the *Life of Apollonius of Tyana.*

Longinus· περὶ υψους Longinus, *On the Sublime.* Wilde's handwritten Greek omits accent and breathing marks for the second word.

Sextus Empiricus Sextus Empiricus (fl. c. 200 A.D.), a doctor of medicine and skeptical philosopher, discussed beauty in his *Pyrrhonistic Sketches.*

* * *

[37]

"dans la vie morale il est beau de quelquefois faire naufrage" "in a righteous life it is sometimes seemly to be shipwrecked."

the ancient Greek sarcophagus of Nicholas of Pisa Probable sources: John Ruskin, "Nicholas the Pisan," in *Val D'Arno* (1873), *The Works of John Ruskin,* ed. E. T. Cook and Alexander Wedderburn, Library Edition, 39 vols. (London: George Allen, 1903–12), XXIII, 14–18; John A. Symonds, *The Fine Arts,* Vol. 3 of *Renaissance in Italy,* 7 vols. (New York: Henry Holt, 1908), 78.

* * *

[39]

the beasts of unrighteousness at Ephesus 1 Cor. 15:32.

are though The word "are" has been written in the margin; the rest of the paragraph is indented. The sentence makes better sense without the "are."

a place where . . . fire is not quenched Mark 9:44; Isa. 66:24.

* * *

[40]

Many of the entries on this page are based on Theodor Mommsen, *The History of Rome,* trans. W. P. Dickson, 4 vols. (London: 1861; rpt. New York: Scribner's 1868), I, 302–6.

The Roman was educated . . . gravitas was the ideal Confirmed source: Mommsen, *History,* I, 305–6. Mommsen does not mention the two Greek ideals.

pater familias father of a family, master of a household

civis citizen

gravitas weight

"ευτραπελια" ready wit, liveliness

φιλοσοφειν μετ' ερωτος to philosophize (to love wisdom) with love. Probable source: Thucydides 2.40. Wilde uses this Greek phrase in "The Portrait of Mr. W. H." (CW, 1195), where the narrator remarks, "How that phrase had stirred me in my Oxford days!"

φιλοκαλειν μετ' ευτελειας to love beauty with frugality. Probable source: Thucydides 2.40. Wilde has changed the passage in Thucydides, which reads, "for we are lovers of beauty with frugality and love wisdom without softness."

* * *

[41]

Roman literature began . . . spirit of each nation Confirmed source: Mommsen, *History,* I, 585.

Mommsen says . . . is this so? Confirmed source: Mommsen, *History,* I, 304.

Camenae Roman goddesses, perhaps water deities, usually identified with the Muses.

Romans had no . . . vates is Keltic Confirmed source, Mommsen, *History,* I, 305.

The centuries during . . . rationalism of Euhemeros Confirmed source: Mommsen, *History,* I, 497–98. Euhemeros (fl. 311–298 B.C.) wrote *Sacred Scripture,* a fictional account of travels containing a rationalist, anthropomorphic theory of the gods' origins in human heroes.

to dramatic art through Euripides Confirmed source: Mommsen, *History,* I, 528.

* * *

[43]

Between the Tarquins . . . his untimely wisdom Confirmed source: Mommsen, *History,* I, 580; 511.

* * *

[47]

Συνοικίσμος coming together like a family; synoecism, combining into one city-state. Wilde misplaces the accent; it should appear over the final omicron. He placed no accent over the second appearance of the word.

* * *

[49]

Roman history comes . . . Tatius, as the [continued on p. 51] Confirmed source: Mommsen, *History,* I, 202–3.

* * *

[50]

Just as they . . . from their law Confirmed source: Mommsen, *History,* I, 217; see also p. 223.

alii . . . orabunt causas melius others will plead cases better. Confirmed source: Vergil, *Aeneid,* 6.847–49.

dare jura to give laws. Possible source: Vergil, *Aeneid,* 6.849–51; Wilde may be paraphrasing Vergil's mission for Rome.

* * *

[51]

[continued from p. 49] **early right of provocatio . . . of the Horatii-** Confirmed source: Mommsen, *History,* I, 203n; see also p. 591. Provocatio can be translated as an appeal (to a higher court of law).

Roman legislation was . . . and immature civilization Confirmed source: Mommsen, *History,* I, 213–14.

✺ ✺ ✺

[52]

jurisconsulti lawyers

πραγματικοι attorneys or statesmen

"Domus jurisconsulti totum oraculum civitatis" "The home of the lawyer the whole place where the oracle of the state (the state's truth) is given." Cicero, *De Oratore,* ed. Augustus S. Wilkins (Oxford: Clarendon Press, 1892), Bk. I, S45.200, p. 184.

"egregie cordatus homo catus AEliu' Sextus" "especially prudent man Catus Aelius Sextus." Cicero, *De Re Publica,* I, 18, 30; *De Oratore,* Bk. I, S45.198, p. 181.

"ad album et rubricas" "to a white table and chapter headings." Quintilian, *Institutio Oratoria,* ed. H. E. Butler, 4 vols. (London: Heinemann; New York: Putnam's, 1921), Bk. XII, S3.11, I, p. 406.

"HUIC MONUMENTO . . . ET JURISCONSULTUS" "FROM THIS MONUMENT MALICE AFORETHOUGHT BE ABSENT AND A LAWYER."

✺ ✺ ✺

[53]

responsa answers

auctoritas pendentum the authority of the ones weighing

jus naturale natural law

φύσει ϑικίον by nature just; Wilde mistakenly accents the second iota in *ϑίκαιον.*

Dare jura to give laws

✺ ✺ ✺

[55]

flights of . . . <u>at all</u> Probable source: John A. Symonds, *Shelley* [English Men of Letters] (London: 1879; rpt. New York: Harper & Bros. 1879), p. 13. Symonds saw in Shelley "that untamed intellectual ambition—that neglect of the immediate and detailed for the transcendental and universal—which was a marked characteristic of his genius, leading him to fly at the highest while he overleaped the facts of ordinary human life."

as are those flights "are" is written over "is."

function of criticism . . . in the Wilderness Probable source: Matthew Arnold, "The Function of Criticism at the Present Time" (1864). Wilde's parenthetical remarks repeat two of Arnold's points, and Wilde appropriates Arnold's closing figure of the wilderness from the Old Testament (with Arnold as Moses leading the Israelites) to the New Testament and John the Baptist. See "The Function of Criticism at the Present Time," *Lectures and Essays in Criticism,* vol. 3 of *The Complete Prose Works of Matthew Arnold,* ed. R. H. Super, 11 vols. (Ann Arbor: U. of Michigan P., 1960–77).

✻ ✻ ✻

[63]

'contra leges naturae' "against the laws of nature"

✻ ✻ ✻

[68]

So Lucretius says . . . been cut off Titus Lucretius Carus, *De Rerum Natura,* ed. H. A. J. Munro, 4th edn., 2 vols. (Cambridge: Deighton Bell, 1886), vv. 1024–27. Wilde might have used an edition like Monro's Latin text, with notes and translation, which first appeared in 1848 and reappeared in four revisions through 1886.

✻ ✻ ✻

[69]

το καλον the beautiful

✻ ✻ ✻

[71]

"confarreatio" a form of marriage ceremony

άροτος seed-time; begetting of children

"surgere coeperam ut ad te redirem" "I begin to rise in order to return to you." St. Augustine, *The Confessions of Augustine,* Bk. 3, Chap. 4.

"de silvestri cacumine patriam pacis" "from the wooded summit the homeland of peace"

✻ ✻ ✻

[72]

Gods of the Indigitamenta Roman household gods of very limited powers and functions.

'Αγνωστω Θεω' "To an Unknown God." Acts 17:23. Wilde omitted accents and subscripts in his handwritten Greek.

Rediculus Taditanus Rediculus, surnamed Tutanus, was the unknown power credited with causing Hannibal's retreat in 211 B.C.

* * *

[74]

now the . . . Roman religion Cf. the "Rise," CW, 1146.

* * *

[75]

When he announced . . . no shock Cf. the "Rise," CW, 1146.

"Quantum religio potest suadere malorum" "How much religion can persuade [men to do] evil." Wilde has changed to "Quantum" and "potest" from Lucretius, 1.101, which reads "Tantum religio potuit suadere malorum," or, So much religion could persuade [men to do] evil.

* * *

[76]

'tauribolium' A religious ritual associated with the cults of Cybele and Venus Caelesta in the pagan, post-Republican Roman Empire (c. A.D. 130–400). Performed on official occasions or for individuals, the rite involved the sacrifice of a bull. Its blood bathed the recipient, who stood in a ditch below.

* * *

[77]

the lonely Emperor at Capri Tiberius

* * *

[85]

Aufklarung enlightenment

* * *

[93]

It is impossible . . . how they lie and move Probable source: John Tyndall, "Professor Virchow and Evolution" (1878), *Fragments of Science: A Series of Detached Essays, Addresses, and Reviews,* 2 vols., 6th edn. (New York: Appleton, 1900), II, 407. Here and elsewhere Tyndall refers to the "two Incomprehensibles"; see pp. 221–22, 391. This essay is also a source for CN, 61.

at certain period of refrigeration Probable source: Tyndall, ''Professor Virchow and Evolution,'' p. 393: ''[W]ould one consequence of the refrigeration of the mass [planet] be the development of organic forms?''

* * *

[94]

''The phantom of Vital Force Possible source: Tyndall, ''Vitality'' (1867) *Fragments,* II, 46. Tyndall rejects Vital Force, but not in this language.

* * *

[95]

Vitalistic conception of life . . . phenomena of animal heat Possible source: Tyndall, ''Science and Man'' (1877), *Fragments,* II, 348. Tyndall mentions that J. R. Mayer's experiments on Animal Heat are vindicated in contrast to notions of ''Vital Force.''

* * *

[100]

of being or of essence the ''or'' is written over ''of.''

* * *

[101]

The opposition between . . . whole and the universal Confirmed source: William Wallace, ''Prolegomena,'' *The Logic of Hegel,* by G. W. F. Hegel, trans. William Wallace (Oxford: Clarendon Press, 1874), p. lix.

In the rhythm . . . the absolute Idea Confirmed source, Wallace, ''Prolegomena,'' pp. lxxi–lxxii. Wilde summarizes, first taking the idea of rhythm and the dialectic from p. lxxi, then adding the progress of philosophy from p. lxxii.

* * *

[105]

The great difficulties . . . by such doctrines Confirmed source: Wallace, ''Prolegomena,'' p. lx.

as preestablished harmony, divine interposition and the like Confirmed source: Wallace, ''Prolegomena,'' p. lviii.

but this onesided . . . the Absolute Idea Confirmed source: Wallace, ''Prolegomena,'' p. lx. Cf. CB, 199, which is also partly based on the same page in Wallace.

νοησις νοησεως: thought of thought or understanding of understanding.
The mind does . . . of a process Confirmed source: Wallace, "Prolegomena," p. lix.

✻ ✻ ✻

[106]

The road of . . . sanctuary of truth Confirmed source: G. W. F. Hegel, quoted in Wallace, "Prolegomena," pp. xxxi–xxxii. Wilde adds the wedding garment and the comparison to entering the Kingdom of God.

✻ ✻ ✻

[107]

The analytic force . . . the feet of God Confirmed source: Wallace, "Prolegomena," p. xxxvii. Wallace quotes Tennyson's "The Passing of Arthur," *The Idylls of the King,* ll. 422–23, which is the source for the image of the golden chain.
it is vain . . . the Absolute Idea Confirmed source: Wallace, "Prolegomena," xli–xliii. Wilde summarizes three long paragraphs linking ideas and phrases from Wallace by means of the golden chain image. Cf. CB, 197, where Wilde's entry is based on some of the same source material.

✻ ✻ ✻

[109]

Kant showed the . . . empiricism of Hume Confirmed source: Wallace, "Prolegomena," p. xxx.
by pointing out . . . of the Absolute Confirmed source: Wallace, "Prolegomena," p. xxx. In his note Wallace directs the reader to the *Logic,* pp. 50–102; this point appears on pp. 72–73.
it remained for Hegel . . . secrets of God Confirmed source: Wallace, "Prolegomena," p. xxxi.
Jacobi Friedrich Heinrich Jacobi (1743–1819). German philosopher and critic of Spinoza, Hume, Kant, and Schelling.
to him the road to true knowledge Wallace uses the road metaphor on p. xxxi.
to the disintegrating method . . . are unravelled Probable source: Wallace, "Prolegomena," p. xxxii.

✻ ✻ ✻

[111]

Socrates and Kant . . . reconquer the World Confirmed source: Wallace, "Prolegomena," p. xxi. Cf. CB, 209, where the same point is noted.

Thought: like Digestion . . . held about it Confirmed source: Benjamin Jowett, "Introduction" to "Parmenides," *The Dialogues of Plato,* IV, 134. Wilde seems to have been attracted to the word "annihilation," which he substitutes for Jowett's "destruction" and "denial." On Hume's denial of cause and effect, cf. CB, 129.

* * *

[119]

Hume explained . . . hydrogen it contains Confirmed source: W. K. Clifford, "The Philosophy of the Pure Sciences," *Lectures and Essays,* ed. Leslie Stephen and Frederick Pollock, 2 vols. (London: Macmillan, 1879), I, 278.

But this . . . right and wrong Confirmed source: Clifford, "On the Scientific Basis of Morals," *Lectures and Essays,* II, 109–12. Wilde summarizes the section introducing "the tribal self."

a savage is . . . trodden on Confirmed source: Clifford, "On the Scientific Basis of Morals," *Lectures and Essays,* II, 111.

* * *

[121]

All the remarks on this page are drawn from Clifford's "On the Scientific Basis of Morals"; see esp. *Lectures and Essays,* II, 111–14, 119–22. The same points are also argued in Clifford's "Right and Wrong: the Scientific Ground of their Distinction," *Lectures and Essays,* II, 168–73.

* * *

[123]

Tylor who sees . . . biological one Too Confirmed source: Clifford, "Body and Mind," *Lectures and Essays,* II, 36.

modern science . . . hydrogen only one Confirmed source: Clifford, "Virchow on the Teaching of Science," *Lectures and Essays,* II, 307.

* * *

[124]

Tacitus says . . . their absence Tacitus, *Annals,* Bk. III, LXXVI, in *Works,* trans. Arthur Murphy (New York: n.p., 1861), p. 99.

Mr. Be[e]sley Edward Spencer Beesly (1897–1916), professor of history at University College, London (1860–89) and editor of *The Positivist Review;* published *Catiline, Clodius, and Tiberius* in 1878.

* * *

[125]

But modern science . . . a moral sentiment Probable source: Clifford, "Right and Wrong: the Scientific Ground of their Distinction," *Lectures and Essays*, II, 139, 145–47. Clifford argues the point at length, and Wilde makes a summary aphorism. The concluding aphorism of this paragraph Wilde added in pencil, changing the period to a colon.

Clifford says . . . any phenomena Confirmed source: Clifford, "The Philosophy of the Pure Sciences," *Lectures and Essays*, I, 281. See also CB, 137, where the same passage in Clifford is quoted at greater length.

Now here he . . . of the other Confirmed source: Clifford, "The Philosophy of the Pure Sciences," *Lectures and Essays, I*, 320. See also pp. 258, 329. Clifford makes a similar point about continuity in "On Theories of the Physical Forces," *Lectures and Essays*, I, 111–14.

* * *

[127]

Wilde's transcriptions from Wordsworth occasionally differ in small particulars from the versions in the Oxford edition, used here as a standard for line numbers: *The Poetical Works of William Wordsworth*, ed. E. de Selincourt and Helen Darbishire, 2nd edn. (London: Oxford UP, 1952–59).

The most ancient . . . fresh and strong "Ode to Duty," l. 56.

The ideal path . . . paved with stars "Dion," ll. 50–51.

He hath overleaped the eternal bars "Dion," l. 53.

One impulse from . . . the sages can "The Tables Turned: An Evening Scene on the Same Subject," ll. 21–24.

We murder to dissect "The Tables Turned," l. 28.

Heaven lies about us in our infancy "Ode: Intimations of Immortality from Recollections of Early Childhood," l. 66.

come forth into . . . be your teacher "The Tables Turned," ll. 15–16.

Two voices are . . . chosen music liberty "Thought of a Briton on the Subjugation of Switzerland," ll. 1–4.

* * *

[129]

Attempt of the . . . dust has failed Possible sources: Clifford, in "Virchow on the Teaching of Science," *Lectures and Essays*, II, 287, 311–17 reports the case for not teaching spontaneous generation; and John Tyndall, in "Spontaneous Generation," *Fragments*, II, 290–334, reports his experiments with the hay bacillus. He also rejected spontaneous generation in "The Belfast Address," *Fragments*, II, 192.

Berkeley annihilated . . . to an end Probable source: Jowett, "Introduction'' to ''Parmenides,'' *The Dialogues of Plato,* IV, 155. Jowett makes this point regarding cause and effect. See also CB, 111, and CB, 129.

* * *

[130]

the cor cordium of antiquity heart of hearts; ''cor cordium'' is the inscription on Shelley's tombstone; Wilde used the phrase in his poem ''Heu Miserande Puer.'' See *Letters,* p. 40.

* * *

[131]

Parmenides . . . put together Confirmed source: Jowett, ''Introduction'' to ''Parmenides,'' *The Dialogues of Plato,* IV, 130.

* * *

[133]

Plato talks . . . (Jowett• 4• 396) Confirmed source: Jowett, ''Introduction'' to ''Sophist,'' *The Dialogues of Plato,* IV, 396, 462.

There are more . . . (Jowett Plato. 4. 155) Confirmed source: Jowett, ''Introduction'' to ''Parmenides,'' *The Dialogues of Plato,* IV, 155.

* * *

[135]

Knowledge as Plato . . . not "by them" Confirmed source: Jowett, ''Introduction'' to ''Theaetetus,'' *The Dialogues of Plato,* IV, 260, 278.

παθημασι: sufferings

συλλογισμω: reasonings

sensation seems to . . . is a "συγκεχυμένον τι" Confirmed source: Jowett, ''Introduction'' to ''Theaetetus,'' *The Dialogues of Plato,* IV, 278, 270.

συγκεχυμενον τι something confused, confounded, poured together, blurred

* * *

[137]

The experience of the race . . . to certain truths Probable source: Clifford, ''The Philosophy of the Pure Sciences,'' *Lectures and Essays,* I, 279–81. Clifford summarizes and quotes from Herbert Spencer's *Principles of Psychology.*

"the accumulated experience . . . five minutes" (Clifford) Confirmed source: Clifford, ''The Philosophy of the Pure Sciences,'' *Lectures and Essays,* I, 281.

We express our . . . apparatus of thought Confirmed source: Clifford, "The Philosophy of the Pure Sciences," *Lectures and Essays,* I, 281.

* * *

[139]

Huxley says . . . of immediate consciousness Confirmed source: Thomas Henry Huxley, *Hume* (New York: Harper & Bros., 1879), pp. 116–17.

now as regards . . . our mental apparatus Confirmed source: Clifford, "The Philosophy of the Pure Sciences," *Lectures and Essays,* I, 278.

Clifford showed that . . . wheel of life Confirmed source: Clifford, "On Theories of the Physical Forces," *Lectures and Essays,* I, 111–14. Cf. CB, 125.

* * *

[140]

Herbert Spencer rests . . . contents for consciousness Wilde summarizes the Spencerian position. Probable source: Herbert Spencer, *First Principles,* 4th edn. (Philadelphia: David McKay, 1880), pp. 80–81, 162–63, 464–65.

* * *

[141]

As regards however . . . (see Clifford 1.154.) Confirmed source: Clifford, "On the Aims and Instruments of Scientific Thought," *Lectures and Essays,* I, 152–53.

Cubic miles "Cubic" has been written over "hun"; perhaps Wilde had begun to write "hundreds."

* * *

[144]

Science and Hellenism• . . . their immediate heritage Possible source: J. A. Symonds, in his conclusion to *Studies of the Greek Poets,* 3rd edn., 2 vols. (London: Adam and Charles Black, 1893), II, 388–415, argues this position in different language.

"whom the hungry generations tread down" Symonds begins his chapter on "The Genius of Greek Art" with a similar adaptation of the second line of the seventh stanza of Keats' "Ode to a Nightingale." *Studies of the Greek Poets,* II, 361.

* * *

[145]

Baudelaire's passionate cry Oh Lord! give me the strength and the courage to contemplate my heart and my body without disgust, from "Un Voyage à Cythere"; Wilde also used the phrase in *De Profundis, Letters,* p. 480. Cf. "The English Renaissance in Art," *The Essays of Oscar Wilde* (New York: H. S. Nichols, 1916), p. 471.

old Hellenic ideal ζήν κατα φυσιν to live by nature; according to the natural order; by custom or law. Wilde's acute accent should be a circumflex.

Philosophy began with . . . in the laboratory Probable source: Huxley, *Hume,* pp. 49–50: "In truth, the laboratory is the fore-court of the temple of philosophy. . . ."

Religion tells us . . . set on edge Jer. 31:29 and Ezek. 18:2.

latest word of . . . we cannot escape Probable source: Clifford, "The Philosophy of the Pure Sciences," *Lectures and Essays,* I, 281. Cf. CB, 137.

* * *

[147]

Flourens Marie-Jean-Pierre Flourens (1794–1867), working under the sponsorship of Georges Cuiver in Paris, conducted experiments on pigeons (1814–22) to determine what physiological changes occurred after portions of their brains were removed.

* * *

[150]

το μεν ωφελιμον καλον το δε κακον αισχρον that which is useful is beautiful, that which is bad is ugly. Probably a quotation from memory of Plato, *Republic* 457B, which has βλαβερον, "harmful," for Wilde's κακον, "bad."

οτι καλον φιλον εστι, το δ'ου καλὸν ου φιλον εστι because it is beautiful it is beloved, but that which is not beautiful is not loved. Confirmed source: *Theognis* 17.

ουχ αμιν τα καλα πρατοις καλα φαινεται ημες for beautiful things at first do not appear beautiful to us. Confirmed source: Theocritus 13.3.

* * *

[153]

The Plastic Spirit of Greek Literature Wilde here summarizes in an opening paragraph the position based on Symonds and Hegel which he develops on the following pages. Symonds' debt to Hegel may be found in *Studies of the Greek Poets,* II, 376: " 'The Greek Spirit,' says Hegel, 'is the plastic artist forming the stone into a work of art.' And this work of art is invariably the image of a man or woman.'' Symonds goes on to defend this thesis in the following chapter.

* * *

[154]

"Stand off . . . a painter might Wilde refers to Euripides' *Hecuba,* 807–8.

Furies of Orestes . . . frankness of detail Symonds includes the passage describing the furies in a long quotation from *Orestes* in *Studies of the Greek Poets,* II, 45. Cf. the ''Rise,'' CW, 1127.

* * *

[155]

Phidias when asked . . . Homer as the answer The three lines of Homer are *Iliad* 1.528–30; the story about Phidias is given in the Scholia on the *Iliad* passage.

We may compare . . . down a mountain Confirmed source: Symonds, *Studies of the Greek Poets,* II, 154. Cf. Wilde's own translation of a chorus from the *Clouds,* CW, 816.

πλαγιαι sideways, slanting, oblique, athwart; the reference is probably to Aristophanes, *Clouds,* 325.

or compare again . . . and in philosophy Confirmed source: Symonds, *Studies of the Greek Poets,* II, 357–58.

cf. also Shelley's . . . Lament of Bion Confirmed source: Symonds, *Studies of the Greek Poets,* II, 279.

* * *

[156]

compare Aristophanes . . . modern times Dante Confirmed source: Symonds, *Studies of the Greek Poets,* II, 189. The scales are in *Frogs,* 800ff.; the mortar is in *Peace,* 228ff. Wilde has added the comparison to the first scene of Aeschylus' *Prometheus Bound.*

Κρατος force, power, strength

Βια violence, might

* * *

[157]

hubsch objectif charmingly dispassionate

mark or external sign Wilde probably refers to Euripides' *Hippolytus,* 925–27.

Marlowe floods . . . vehicle of passion Confirmed source: Symonds, *Studies of the Greek Poets,* II, 358–59.

the Faun of Praxiteles Possible source: Symonds, *Studies of the Greek Poets,* II, 375.

* * *

[158]

So the plastic arts . . . vases are in art Confirmed source: Symonds, *Studies of the Greek Poets,* II, 245.

* * *

[159]

the figures on the Medicean tomb Probable source: Symonds, *Studies of the Greek Poets,* II, 391.

* * *

[161]

το απειρον the boundless

ερως τωναδυνατων love of impossible things

music the essentially . . . within corporeal limits Confirmed source: Symonds, *Studies of the Greek Poets,* II, 258–59.

Modern art appeals . . . the external world Confirmed source: Symonds, *Studies of the Greek Poets,* II, 261.

* * *

[163]

Art then may . . . of its nature Confirmed source: Walter Pater, "Winckelmann," *The Westminster Review,* 87 (January–April 1867), 50.

So to the Egyptians . . . its own beauty Confirmed source: Hegel, quoted by Pater, "Winckelmann," 44.

to the Greeks . . . to modern life Confirmed source: Pater, "Winckelmann," 50. Wilde makes a significant change by substituting "the sense of the infinite, and the scientific spirit of criticism" for Pater's "the sense of freedom."

* * *

[165]

The bridge between . . . of Greek tragedy Confirmed source: Pater, "Winckelmann," 47; Pater gives Hegel's *Aesthetik,* 2.2.3, as the source for his remark.

and also in . . . pale mediaeval painter Confirmed source: Pater, "Winckelmann," 48; Pater introduces his remarks as "what Hegel has so cunningly detected" and gives Hegel's *Aesthetik,* 2.3.2, as his source.

* * *

[167]

Of the quality . . . the Dionysiac Theatre Wilde may be replying to Symonds, *Studies of the Greek Poets,* II, 132–33, 135–36, who saw sculpture and tragedy as allied in their "stationary character" and perfectly harmonious effect. Symonds compared the expression of passion in Greek tragedy to "a stationary and prolonged aria" in the opera—a far cry from Dionysiac passion.

* * *

[168]

in the colourless abstractions of a spiritual world Confirmed source: Pater, "Winckelmann," 47–48. Wilde's short meditation on metaphysics is based also on other extracts from "Winckelmann" which appear on the facing page of the notebook, CB, 169.

nomina into numina names into gods, divinities

* * *

[169]

In Raffaelle and . . . it's tears (Hegel) Confirmed source: Pater, "Winckelmann," 48.

"We must renounce . . . contrasts of Life•" Confirmed source: Pater, "Winckelmann," 50.

* * *

[171]

Necessity is not . . . of the World Confirmed source: Pater, "Winckelmann," 50.

* * *

[172A—recto]

The end of life . . . through the means Confirmed source: Walter Pater, "On Wordsworth," *The Fortnightly Review,* n.s. 15 (April 1874), 465. Phrases which appear on this page are also found in "The Critic as Artist," CW 1023, 1039, 1041–42, and in *De Profundis,* in *Letters,* pp. 476, 505.

* * *

[172A—verso]

Those who object . . . therefore strictly relative Cf. CB, 151.

* * *

[173]

The thoughts of men are widened with the process of the suns Confirmed source: Tennyson, "Locksley Hall," l. 138.

Plastic Art can . . . desire of escape Used in "The English Renaissance of Art," *The Essays of Oscar Wilde,* pp. 448–49.

* * *

[175]

"radios nullos sive . . . societatem humanam diffundat" Confirmed source: Francis Bacon, *De Augmentis Scientiarium,* VII.i, in *The Works of Francis Bacon,* ed. James Spedding, Robert L. Ellis, and Douglas D. Heath, 14 vols. (London: Longmans, 1857–74), I, 719. The sentence from which Wilde draws the Latin phrase is translated in Spedding, Ellis, and Heath as: "But for mere contemplation which should be finished in itself without casting beams of heat and light upon society, assuredly divinity knows it not" (V, 8).

mystery in the world Wilde corrected from "mysteries" to "mystery" by writing "y" over "ies."

"versamur adhuc plane in atriis naturae" "we pass time thus far entirely in the houses of nature"

"neque adhuc Experientia literata facta est" Confirmed source: Francis Bacon, *Novum Organum* I.101, *The Works of Francis Bacon,* I, 204. The phrase is translated as: "and experience has not yet learned her letters" (IV, 96).

* * *

[177]

"Sunt enim non . . . eremi et vastitates" Confirmed source: Bacon, *Novum Organum,* I.78, *The Works of Francis Bacon,* I, 186. translated as: "In times no less than in regions there are wastes and deserts" (IV, 77).

"subtilitas naturae . . . multis partibus superat" Confirmed source: Bacon, *Novum Organum,* I.10, *The Works of Francis Bacon,* I, 158, translated as:

"The subtlety of nature is greater many times over than the subtlety of the senses and understanding" (IV, 48).

Scientia et potentia humana in idem coincidunt Confirmed source: Bacon, *Novum Organum,* I.3, *The Works of Francis Bacon,* I, 157, translated as: "Human knowledge and human power meet in one" (IV, 47).

ipsissimae res veritas et utilitas Confirmed source: Bacon, *Novum Organum,* I.124, *The Works of Francis Bacon,* I, 218. Wilde has shortened Bacon's phrasing: "Itaque ipsissimae res sunt (in hoc genere) veritas et utilitas," which is translated as: "Truth therefore and utility are here the very same things" (IV, 10).

* * *

[179]

To Shelley the bees . . . Nurslings of Immortality Confirmed source: P. B. Shelley, *Prometheus Unbound,* I.745–49. Symonds also admired this passage, and Wilde may have found it quoted in his *Shelley,* p. 125, or in *Studies of the Greek Poets,* II, 261, immediately following the passage which Wilde copied on CB, 161.

* * *

[181]

That we have . . . the arteries pinched Confirmed source: Clifford, "Right and Wrong: the Scientific Ground of their Distinction," *Lectures and Essays,* II, 161.

* * *

[183]

Κοσμος cosmos

* * *

[185]

"Rien n'est vrai que le beau" "Nothing is true but the beautiful"

"Rien n'est vrai . . . simple and absolute Confirmed source: Algernon Charles Swinburne, "Notes on Some Pictures of 1868," *Essays and Studies* (London: 1875; rpt. Chatto & Windus, 1911), pp. 379–80.

As it is . . . the noblest education Wilde made a similar observation about Plato and education in "The English Renaissance of Art," *The Essays of Oscar Wilde,* pp. 474–75.

ευθυς παιζοι εν τοις καλοις: immediately he could play among beautiful things. Confirmed source: Plato, *Republic,* 558B; Wilde has added the article ***τοις*** to Plato's text.

δεῖ γαρ τελευταν τα μουσικα εις τα του καλου ερωτικα for music must result in the love of beauty. Confirmed source: Plato, *Republic,* 403C. Again, Wilde has slightly changed Plato's δέ που "but somehow," to γὰρ, "for."

La beauté est . . . pas a *[sic]* demi beauty is perfect, beauty can do all things, beauty is the only thing in the world that doesn't exist by halves. Confirmed source: Swinburne, "Notes on Some Pictures of 1868," *Essays and Studies,* p. 380.

* * *

[187]

Clifford accounts for . . . from three planes Confirmed source: Clifford, "The Philosophy of the Pure Sciences," *Lectures and Essays,* I, 339–40.

* * *

[189]

If Philosophy aims . . . begins it's flight Confirmed source: G. W. F. Hegel, *Philosophie des Rechts,* p. 21, quoted in W. Wallace, "Prolegomena," *The Logic of Hegel,* p. xxiv. Wilde quotes and summarizes from Wallace's commentary and from the passage of Hegel which Wallace reproduces.

Σοφια knowledge

"η μεν γαρ σοφια ουδεν θεωρει εξ ων εσται ευδαιμων ανθρωπος". (Eth. 6. 13) for wisdom does not consider because of what things a man will be happy. Wilde changes Aristotle's future θεωρήσει, "will not consider." Wilde refers to the second sentence in Bk. VI, Chap. 13, of Aristotle's *Ethics,* which considers the relationship between virtue and wisdom. In some editions and translations the passage appears as the second sentence of Chap. 12.

αρετη virtue

it is good . . . for its existence This idea appears in "The Critic as Artist," CW, 1039.

Bacon's scornful words . . . bears no fruit Wilde refers to Bacon's remark in *De Augmentis Scientiarum,* III.5, "Nam *Causarum Finalium* inquisitio sterilis est, et tanquam virgo Deo consecrata nihil parit," *The Works of Francis Bacon,* I, 571, translated as: "For the inquisition of Final Causes is barren, and like a virgin consecrated to God produces nothing" (IV, 365).

* * *

[191]

an αισθητον · a τοδε τι a perceived, a something

τοδε τι · που· και νυν something anywhere and now

αει και πανταχου ever and everywhere

* * *

[192]

επακτικοι λογοι inductive reasoning
η μαιευτικη τεχνη the midwifery skill
το οριζεσθαι καθολου to determine in general
τελεολογια teleology

* * *

[193]

τελος end, purpose
διαλεκτική dialectic
Ερως eros, love
Ερως and διαλεκτικη love and dialectic
ιδεα του αγαθου idea, form of the good

In discussion "we . . . embrace to wrestle" Confirmed source: Winwood Reade, quoted by W. K. Clifford in "The Philosophy of the Pure Sciences," *Lectures and Essays,* I, 297: " 'that fruitful faculty [criticism] . . . which embraces to wrestle, and wrestles to embrace.' "

* * *

[195]

In the Parmenides . . . relation to contraries Possible sources: Jowett's "Introduction" to "Sophist," *The Dialogues of Plato,* IV, 384 or 402. Wilde may be summarizing.

In Sophist . . . what it is not Possible sources: Jowett's introductions to "Parmenides" and "Sophist," *The Dialogues of Plato,* IV, 135, 384, 399, 402.

* * *

[197]

Κοσμος cosmos

The condemnation which . . . explain him. (Hegel) Confirmed source: Wallace, "Prolegomena," *The Logic of Hegel,* p. xiii. This is Wallace's opening sentence.

"A Knight of the Holy Ghost" Confirmed source: Wallace, "Prolegomena," *The Logic of Hegel,* p. xxv. Wallace uses Heine's phrase to describe Hegel, but Wilde changes the referent to Plato.

A synthesis of . . . of pure thought Confirmed source: Wallace, "Prolegomena," *The Logic of Hegel,* p. xli.

before dividing into . . . medium of thought Confirmed source: Wallace, "Prolegomena," *The Logic of Hegel,* p. xliii.

The totality of . . . the spirit remains Confirmed source: Wallace, "Prolegomena," *The Logic of Hegel,* p. clxxiv.

* * *

[198]

Kant shattered . . . "Ding-an-sich" Confirmed source: Wallace, "Prolegomena," *The Logic of Hegel,* pp. xlvii, cxv.

Hegel solved the . . . into Logic Confirmed source: Wallace, "Prolegomena," *The Logic of Hegel,* p. cxv; cf. p. xlviii.

The absolute presents . . . the true absolute Confirmed source: Wallace, "Prolegomena," *The Logic of Hegel,* p. cxiii.

Being. Relative Being . . . particular. Individual Confirmed source: Wallace, "Prolegomena," *The Logic of Hegel,* p. cxxxix.

* * *

[199]

Kant annihilated metaphysics Confirmed source: Wallace, "Prolegomena," *The Logic of Hegel,* p. lii.

but he acted . . . men for themselves Confirmed source: Wallace, "Prolegomena," *The Logic of Hegel,* p. lx. Cf. CB, 105, where Wilde makes notes from the same passage in Wallace.

* * *

[201]

Aristotle said . . . exercises itself Confirmed source: Aristotle, *De Anima,* III.4, quoted in Wallace, "Prolegomena," *The Logic of Hegel,* p. liii.

Hegel remarked . . . a substance Confirmed source: Wallace, "Prolegomena," *The Logic of Hegel,* pp. lv–lvii.

Aristotle rem. [arked] that . . . for all things Confirmed source: Wallace, "Prolegomena," *The Logic of Hegel,* pp. cxxv–vi.

what we think . . . idealism of reason Confirmed source: Wallace, "Prolegomena," *The Logic of Hegel,* p. cxlvii.

* * *

[203]

Progress of . . . Quantity. Measure Confirmed source: Wallace, "Prolegomena," *The Logic of Hegel,* pp. cxix ff. ("the Three Parts of Logic").

To reduce quality . . . "one" and "many" Confirmed source: Wallace, "Prolegomena," *The Logic of Hegel,* p. cxli.

ιδια αισθησις private perception
κοινα αισθητα common perception
Quality: Being . . . Measure. Anaxagoras Confirmed source: Wallace, "Prolegomena," *The Logic of Hegel,* p. cxlv.

* * *

[204]

Hegelian dialectic is . . . world of thought Confirmed sources: Wallace, "Prolegomena," *The Logic of Hegel,* pp. cii, clxxx; Jowett, in his introductory matter to Plato, says, "The struggle for existence is not confined to the animals, but appears in the kingdom of thought" (*The Dialogues of Plato,* IV, 406).
ερως eros, love
νικη victory
so ερως and . . . the Platonic ερως Confirmed source: Jowett, "Introduction" to "Symposium," *The Dialogues of Plato,* II, 12–13.
ερως· is the beginning . . . thirst after righteousness Confirmed source: Jowett, "Introduction" to "Symposium," *The Dialogues of Plato,* II, 18.
fruitio Dei enjoyment of God

* * *

[205]

The Platonic ερως . . . inquietude poussante of Leibnitz Confirmed source: Wallace, "Prolegomena," *The Logic of Hegel,* pp. c–ci.
(Symposium) Plato, *Symposium,* 203B–C.
Philosophy began in . . . Not-Being is not Confirmed source: Hegel, *The Logic of Hegel,* trans. W. Wallace, p. 137.
The earliest systems . . . least in them Confirmed source: Hegel, *The Logic of Hegel,* trans. W. Wallace, p. 136.
Modern science is . . . science is geometry Confirmed source: Wallace, "Prolegomena," *The Logic of Hegel,* p. cxlix.

* * *

[206]

The cognition . . . be sensuous Possible source: Hegel, "The Idea of Artistic Beauty, or the Ideal," *Hegel on the Arts* (New York: Ungar, 1979), p. 5: "while it is itself sensuous, art is somehow nevertheless able to deliver man from the power of sensuousness."

* * *

[207]

Lobatchewski was to Euclid . . . Copernicus to Ptolemy Confirmed source: Clifford, "The Philosophy of the Pure Sciences," *Lectures and Essays,* I, 297.

versatility Cf. CN, 23.

"προς απαν συνετον επι παν αργον·" Thucydides 3.82.4. This difficult phrase is translated by Gomme as "an intelligence which can grasp the whole amounted to inactivity in everything."

"tabula rasa" "blank tablet"

* * *

[209]

Kant like Socrates . . . conquer the world Confirmed source: Wallace, "Prolegomena," *The Logic of Hegel,* p. xxi. Cf. CB, 111, where the same point is also noted.

* * *

[213]

"non moins sterile que puerile" "not less sterile than childish"

"poser des questions . . . qui est insondable" "to pose questions which are insoluble, and to fathom the absolute which is unfathomable"

* * *

[214]

Positivism may be described as catholicism without Christianity Confirmed source: T. H. Huxley, "On the Physical Basis of Life" (1868), rpt. in *Lay Sermons, Essays and Reviews* (New York: Appleton, 1878), p. 140. Huxley used the phrase again in "The Scientific Aspects of Positivism" (1869), *Lay Sermons, Addresses and Reviews,* pp. 152–53.

* * *

[215A]

scribere facienda to write things to be done
facere scribenda to make things to be written

7

Notes to Notebook Kept at Oxford

✵ ✵ ✵

—

[inside front cover]

Leroux de' *[sic]* l'Humanité . . . Storia della Critica These sources are cited by Robert Flint in *The Philosophy of History in France and Germany* (Edinburgh and London: Blackwood and Sons, 1874), pp. 25, 30n.

—

[facing inside front cover]

Acusilaus Probable source: William Mure, *A Critical History of the Language and Literature of Antient Greece,* 2nd edn., 5 vols. (London: Longman, Brown, Green, Longmans, and Roberts, 1854–59), IV, 138–39. This was a standard reference for Oxford undergraduates in the *Literae Humaniores* program.

✵ ✵ ✵

[1]

Survival of culture . . . Key and Bible Confirmed source: Edward B. Tylor, *Primitive Culture: Researches into the Development of Mythology, Philosophy, Religion, Language, Art, and Custom,* 7th edn. (New York: Brentano's, 1924), p. 75.

"le petit bon homme vit encore" "the little good man lives again"

✵ ✵ ✵

[2]

in history . . . Evolutions Cf. the "Rise," CW, 1125; see also "The English Renaissance of Art," *The Essays of Oscar Wilde,* p. 446.

* * *

[3]

Nature as Aristotle said . . . general laws Confirmed source: Tylor, *Primitive Culture,* pp. 2–5; cf. the "Rise," CW, 1120, where Wilde says the basis of Aristotle's philosophy is that "Nature maketh no thing in vain"; see also CW, 1113, 1122, 1124, 1126.

* * *

[5]

Spiritualism an instance of survival becoming revival Confirmed source: Tylor, *Primitive Culture,* p. 17.

Prehistoric Archaeology . . . wooly rhinoceros Confirmed source: Tylor, *Primitive Culture,* p. 58.

Boucher de Perthes' Jacques Boucher de Crèvecoeur de Perthes (1788–1868), French archeologist, studied paleolithic tools and fauna at Abbeville, France, in 1836. His findings provided evidence that humans existed long before biblical history said they were created. See Marvin Harris, *The Rise of Anthropological Theory: A History of Theories of Culture* (New York: Thomas Y. Crowell, 1968), p. 147.

Lucretius v. 1281• quite accurate Confirmed source: Tylor, *Primitive Culture,* pp. 60–61; cf. the "Rise," CW, 1122, 1146.

* * *

[7 (and 6)]

To adjust . . . the historian Confirmed source: William E. H. Lecky, *History of the Rise and Influence of the Spirit of Rationalism in Europe,* rev. edn., 2 vols. (New York: Appleton, 1914), I, 14–15. Hegel's discussion of great men in history may also be a source. See his *The Philosophy of History,* trans. J. Sibree (New York: P. F. Collier & Son, 1901), pp. 76–77.

Aristotles dictum . . . "εκ μίκρων αλλ' ου περὶ μίκρων." In the "Rise" (CW, 1136) Wilde translates it as "not about trivial issues but arising from trivial causes." The Loeb edition translates it as "Factions arise not about but out of small matters." Confirmed source: Aristotle, *Politics,* V, Chap. 3, 1. 1. Wilde incorrectly accents the iotas in *μικρων* and omits other accents.

* * *

[8]

appears also . . . philosophy of history Confirmed source: Auguste Comte, *Système de Politique Positive; ou, Traité de Sociologie, Instituant la Religion de l'Humanité,* 4 vols. (Paris: Carilian-Goeury et Vor Dalmont, 1851–54), III, 540–41.

* * *

[9]

R• change in the conception . . . belong to history Confirmed source: H. T. Buckle, *History of Civilization in England,* 2 vols. in 4 pts. (London: J. W. Parker & Son, 1857 and 1861), I, ii, 415.

the statical element . . . physical phenomena Cf. the "Rise," CW, 1122.

Cousin said . . . deduce the race Confirmed source: Victor Cousin, "Cours de l'Histoire de la Philosophie," *Oeuvres de Victor Cousin,* (Bruxelles: Société Belge de Libraire, 1840), I, 63. Cousin (1792–1867) was a French philosopher whose theory put him in the camp of "materialists" like George Grote, who believed that human history was mainly determined by such "external factors" as geography and climate. Wilde quotes this remark of Cousin's in the "Rise," CW, 1127n.

* * *

[10]

if the hammer . . . be unnecessary Confirmed source: Aristotle, *Politics,* I, 4, 1253b, ll. 38–40. Cf. Wilde's remarks on slavery in "The Soul of Man Under Socialism," CW, 1089.

* * *

[11]

Even animal monstrosities . . . of crystals Confirmed source: Buckle, *History,* I, ii, 653n, 657–58; II, ii, 444, 444n.

Pinels book P. Pinel, *Traité Médico-Philosophique sur l'Aliénation Mentale* (Paris: J. A. Brossan, 1809).

it is impossible . . . Luther or Voltaire Confirmed source: Lecky, *History,* rev. edn., I, 8.

* * *

[13]

Humanitarian sympathy . . . a particular capital Confirmed source: Lecky, *History,* I, 309–10.

* * *

[15]

How Aristotle . . . with heresy Confirmed source: Lecky, *History,* I, 380n.

* * *

[16]

the chief moral . . . human dignity Possible source: Edward Gibbon, *The History of the Decline and Fall of the Roman Empire,* 6 vols. (New York: Harper and Bros., 1851), I, 11. This idea was widely believed and often stated and could have come from other sources.

* * *

[17]

What is morality . . . which is death Possible source: Herbert Spencer, *The Data of Ethics* (New York: A. L. Burt, 1879), pp. 19–22, 51, 117, 177. This comment summarizes Spencer's general definition of life as "the continuous adjustment of internal relations to external relations" (p. 21) and so is used in many places. Wilde here applies it specifically to the evolution of ethics.

in the ancient . . . Xtian martyrs Confirmed source: Lecky, *History,* II, 102, 102n.

* * *

[19]

in early civilizations . . . of the bayonet Confirmed source: Lecky, *History,* II, 205–7.

* * *

[21]

The inductive method . . . into Science Confirmed source: Flint, p. 25; cf. the "Rise," CW, 1126.

Philosophy of History . . . contained in them: Confirmed source: Flint, *Philosophy,* p. 8; cf. the "Rise," CW, 1126.

History may be . . . only scientific one Cf. the "Rise," CW, 1126; also cf. the remark on revolutions on p. 2 of this notebook.

τέλος purpose

* * *

[22]

see on Shelleys . . . at the orrery Confirmed source: J. A. Symonds, *Shelley* (New York: Harper and Bros., 1894), pp. 14–15. This volume in the English Men of Letters series was first published in 1878. An orrery is an apparatus showing the relative positions and motions of bodies in the solar system by means of balls moved by a wheelwork. It was named for Charles Boyle, fourth Earl of Orrery.

also Leonardo . . . had been differentiated Possible source: Comte, *Système,* III, 539–40.

so also Goethe . . . pupil of science Possible source: John Tyndall, "Scientific Use of the Imagination," *Fragments of Science: A Series of Detached Essays, Addresses, and Reviews,* 6th edn., 2 vols. (New York: Appleton, 1900), II, 131–33; "The Belfast Address," II, 148–49, 158–59.

* * *

[23]

versatile See CB, 207, on versatility.

* * *

[25]

heredity as in . . . Evolution and progress Possible source: J. A. Symonds, *Studies of the Greek Poets,* 3rd edn., 2 vols. (London: Adam and Charles Black, 1893), II, 142n. *The Spanish Gypsy* (1868) is a verse drama by George Eliot. See Gordon Haight, *George Eliot: A Biography* (Oxford: Clarendon, 1968), p. 376, for a discussion of Eliot on heredity.

N. how science and democracy join hands Possible source: Buckle, *History,* I, ii, 662. There Buckle writes: "The hall of science is the temple of democracy." This idea was held, however, by many people.

* * *

[26]

In criticising Machiavelli . . . Econ. does his Possible source: Hegel, *The Philosophy of History,* trans. J. Sibree (New York: P. F. Collier & Son, 1901), p. 506. Hegel characterizes Machiavelli's principles as "the principles on which alone states could be founded in the circumstances of the times."

Social effects . . . inductive philosophy? This topic is discussed in Buckle, *History,* II, ii, 455ff.

Rem• however . . . not inductive Confirmed source: Buckle, *History,* II, ii, 341–65.

* * *

[27]

Aristotle knew nothing . . . theory of averages Confirmed source: Buckle, *History,* II, ii, 459; see, for Aristotle, I, i, 123ff.; cf. the "Rise," CW, 1148.

The rise of . . . middle trading classes Confirmed source: Buckle, *History,* II, ii, 457.

a Deductive system . . . and ancient Greece Confirmed source: Buckle, *History,* II, ii, 322ff., 456, 459. On II, ii, 337, Buckle says that both induction and deduction are required to arrive at scientific truth. Induction without deduction, he claims, leads to mere empiricism, not science. This notion shaped much of Wilde's thinking about historical criticism. Buckle saw induction as analytic, deduction as synthetic. See II, ii, 329.

Liebniz *[sic]* however said . . . want of ideas This is certainly Leibniz's position, but Wilde may have rewritten it into an aphoristic form, since it is difficult to find stated precisely this way. Possible source: *New Essays on Human Understanding,* trans. and ed. Peter Remnant and Jonathan Bennet (Cambridge: Cambridge UP, 1981), p. 417 (Bk. IV, Chap. 7, §11): "It is a different matter when the truths . . . are always or at least usually true, which are learned by observation or grounded in experiments, and for which there are no wholly convincing reasons. But these truths are beside the point here, since they are not known through the connection of ideas." Wilde quotes the remark again on p. 44 of this notebook.

* * *

[28]

nothing can dispense . . . moral aim Probable source: Buckle, *History,* II, ii, 469; cf. the "Rise," CW, 1115.

nothing is easier . . . to use them Possible source: Buckle, *History,* I, i, 3.

the same artistic . . . sphere of History See the "Rise," CW, 1131, for a revised form of this sentence.

* * *

[29]

For a scientific . . . of the future Possible source: Buckle, *History,* I, i, 6–15; cf. the "Rise," CW, 1122.

In the character . . . still holds place Confirmed source: Buckle, *History,* II, ii, 469–71; cf. the "Rise," CW, 1125.

* * *

[30]

Plato underrated induction . . . Aristotle and Newton Confirmed source: Buckle, *History,* II, ii, 450.

Hunter and Cullen . . . rather than physiological Confirmed source: Buckle, *History,* II, ii, 420–52; cf. the "Rise," CW, 1136. William Cullen (1710–1790) was professor of medicine at Edinburgh. William Hunter (1718–1783) was a student of Cullen's and later professor of anatomy at the Royal Academy.

* * *

[31]

The microscope has . . . and unhealthy condition Confirmed source: Buckle, *History,* II, ii, 417–19. Cf. the "Rise" (CW, 1136), where Wilde discusses the relation of pathology and physiology to the development of a scientific method. There he seems to rely on the following idea in Buckle: "pathology should be investigated deductively rather than inductively, and that morbid anatomy and clinical observations may verify the conclusions of science, but can never supply the means of creating the science itself" (I, ii, 364n.).

* * *

[33]

Adam Smith . . . the abstract method Confirmed source: Buckle, *History,* II, ii, 340–41.

* * *

[35]

Asceticism opposed by . . . to commercial interests Confirmed source: Lecky, *History,* II, 350–52.

Ancient colonies due . . . Vasco di *[sic]* Gama Confirmed source: Lecky, *History,* II, 351–52, 352n.

* * *

[36]

persistence of force This Spencerian formulation appears in *First Principles* (Philadelphia: David McKay, 1880), e.g., p. 454. Cf. p. 37 of this notebook.

Matter[,] motion and force . . . unaccountable and unnatural Confirmed source: Spencer, *First Principles,* pp. 468–70.

* * *

[37]

Three primary truths . . . the last ultimate Confirmed source: Spencer, *First Principles,* p. 454; cf. the "Rise," CW, 1124.

However: the interpretation . . . a symbol still Confirmed source: Spencer, *First Principles,* p. 470; cf. the "Rise," CW, 1144.

✵ ✵ ✵

[39]

Rem. what every . . . he follows Cf. the "Rise," CW, 1135.
Novum organum new instrument
οργανον *organon,* instrument
So in judging . . . which is valueable *[sic]* Cf. the "Rise," CW, 1125–28ff.

✵ ✵ ✵

[40]

rem• a wonderful . . . Greeks and barbarians Possible source: Herodotus, *The Histories* (New York: Penguin, 1954), p. 61.

✵ ✵ ✵

[41]

Rem. Xenophon's tract . . . the Athenian democracy Confirmed source: Carel Gabriel Cobet, *Commentatio qua continetur Prosopographia Xenophontea in Certamine Literario Civium* (Lugduni Batavorum: J. Luchtmans, 1836), p. 29; Adolf Kirchoff, *Über die Schrift vom Staate der Athener* (Berlin: Akademie der Wissenschaften, 1874), pp. 1–2.

✵ ✵ ✵

[42]

the poetic imagination . . . the deductive method Confirmed source: Buckle, *History,* II, ii, 395–96.
So Ruskin shows . . . aegis on her breast This quotation runs over to p. 44. Confirmed source: John Ruskin, "The Queen of the Air," in *The Works of John Ruskin,* ed. E. T. Cook and A. Wedderburn, Library edn., 39 vols. (London: George Allen; New York: Longmans, Green, 1903–12), XIX, 292. As Cook and Wedderburn note, in this passage Ruskin refers to a lecture by John Tyndall which he heard at the Royal Institution on January 15, 1869. The lecture, "On Chemical Rays, and the Light of the Sky," was printed in the *Notices of Proceedings,* V, pp. 429–50. On the blue of the sky as reflected light, see p. 440. In their introduction to this study of Greek mythology the editors discuss Ruskin's intellectual relation to Max Müller:

> In considering "The Queen of the Air" as a contribution to the study of Greek mythology, the reader should remember the date at which it was written. The

views of the philological school, headed by Max Müller, were then in the ascendant. "Comparative Mythology," as the philological school understood the term, consisted of a comparison of the roots of words; mythology was a "disease of language," and the common origin of all the myths was to be found in natural, and especially in solar, phenomena. This is the doctrine to which, by implication, Ruskin assented; indeed, he expressly refers, in his Preface (p. 292), to "the splendid investigation of recent philologists," and "it is the task of the Philologists," he says (§1, 296), "to account for the errors of antiquity" (XIX, lxvii).

* * *

[43]

Rem• how the . . . modern scientific truths Cf. the "Rise," CW, 1122. See also CB, 183, on mysticism.

the problem really . . . earth and particulars Confirmed source: Buckle, *History,* II, ii, 395–96.

* * *

[44]

So also on . . . as a fly Confirmed source: Ruskin, "The Queen of the Air," §34–35, *The Works of John Ruskin,* XIX, pp. 331–32. The reference is presumably to *The Iliad,* 17.570.

The attaching too . . . want of ideas Leibnitz, previously quoted on page 27 of this notebook.

* * *

[45]

Buckle points out . . . this elevating character Confirmed source: Buckle, *History,* II, ii, 397.

Two physical philosophers . . . on Heat Confirmed source: Buckle, II, ii, 399, 390ff. See also Wilde's remark about "heat and its laws," in the "Rise," CW, 1122n.

Joseph Black Joseph Black (1728–1799), Scottish chemist and physicist. "Moreover, his attention was engaged on studies which ultimately led to his doctrine of latent heat. He noticed that when ice melts it takes up a quantity of heat without undergoing any change of temperature, and he argued that this heat, which as was usual in his time he looked upon as a subtle fluid, must have combined with the particles of ice and thus become latent in its substance. This hypothesis he verified quantitatively by experiments, performed at the end of 1761." *The Encyclopaedia Britannica,* 11th edn., 29 vols. (Cambridge, Eng.: Cambridge University Press, 1910–11), IV, 18–19.

Sir John Leslie Sir John Leslie (1766–1832), Scottish mathematician and physicist, wrote *An Experimental Enquiry into the Nature and Properties of Heat.*

"He compared the emissive, absorbtive *[sic]* and reflecting powers of various substances to radiant heat. This work was carried out with the help of his 'differential thermometer' and a 'Leslie cube.'" *The Encyclopaedia Brittanica,* XVI, 492.

* * *

[47]

Distinction between organic . . . we call inorganic matter Confirmed source: Buckle, *History,* II, ii, 418–19, 418n. Note, however, that in the last sentence Wilde adds a Hegelian hypothesis to Buckle's observation.

* * *

[49]

Hume's argument . . . often verified experience Confirmed source: Thomas Henry Huxley, *Hume,* English Men of Letters series (New York: Harper & Bros., 1879), pp. 128–29.

Hume's argument See David Hume, *An Inquiry Concerning Human Understanding* (Indianapolis: Bobbs-Merrill, 1955), sect. 10, pt. ii, for his discussion of miracles.

Upon the other hand . . . question of Evidence Confirmed source: Huxley, *Hume*, pp. 131ff.

To define . . . miracle is one Confirmed source: Huxley, *Hume,* p. 128.

* * *

[51]

Comparative anatomy shows . . . anatomy of the mind Confirmed source: Huxley, *Hume,* p. 105. Huxley's main point, which precedes this entry, is rather important: "In short, it seems hard to assign any good reason for denying to the higher animals any mental state, or process, in which the employment of the vocal or visual symbols of which language is composed is not involved; and comparative psychology confirms the position in relation to the rest of the animal world assigned to man by comparative anatomy" (p. 104). Compare John Tyndall's remarks "on tracing the line of life backwards" in "The Belfast Address," *Fragments of Science,* II, 190–91.

* * *

[53]

Force is always . . . on ~~experience~~ opinion Confirmed source: David Hume, "On the First Principles of Government," *Essays Moral and Political* (Edinburgh: R. Fleming & A. Alison, 1741), which Huxley quotes on p. 15.

the social organism . . . so may social institutions Confirmed source: Huxley, *Hume,* pp. 19–20.

all knowledged *[sic]* is . . . experience constitute knowledge Confirmed source: Huxley, *Hume,* pp. 62–66, 86n.

* * *

[55]

object of philosophy . . . science and psychology Confirmed source: Huxley, *Hume,* pp. 46–48.

While psychology rests . . . function of the brain Confirmed source: Huxley, *Hume,* pp. 77–78. Tyndall also uses the analogy of the telegraph operator in "The Belfast Address," *Fragments of Science*, II, 165. Wilde mentions the telegraph operator again on p. 63 of this notebook.

* * *

[56]

is the voice . . . locusts and wild honey? Matt. 3:3–4; Isa. 40:3. See "The English Renaissance of Art," *The Essays of Oscar Wilde,* p. 471; "The Critic as Artist," CW, 1043.

* * *

[57]

matter is what . . . harmony: clock metaphor Confirmed source: Huxley, *Hume,* pp. 163–64.

Occasionalists A group of Dutch followers of Descartes who lived during the seventeenth century, they were interested in the problem of the relation of mind and body, believing the two were distinct. The will does not move bodies and physical occurrences do not produce ideas in us: they are only occasional causes for God's producing them in us. Mental and physical processes run parallel to each other, a parallelism caused by the continual interference of God. See Frank Thilly and Ledger Wood, *A History of Philosophy,* 3rd edn. (New York: Holt, Rinehart and Winston, 1957), pp. 315–16.

Leibnitz In the *Monadology,* sect. 64, Leibnitz explains that the relationship of soul and body is a harmony preestablished by God. There is parallelism but no interaction of the two, and God does not intervene. Organic bodies are "divine machines." See Thilly and Wood, *A History of Philosophy,* pp. 390–91.

* * *

[59]

(3) the Berkeleyan hypothesis . . . thought and extension Confirmed source: Huxley, *Hume,* p. 164. Berkeley inverts Lockean empiricism to establish idealism and refute materialism. "If as Locke claimed, the basis of our knowledge is sensation and reflection, and we know only ideas, how can we know a world of bodies, a material world outside us? . . . The belief in matter, therefore, leads to atheism and materialism. . . . We can avoid these implications only by getting rid of the premise from which they spring—the assertion that matter exists" (Thilly and Wood, *A History of Philosophy*, pp. 358–59).

Spinoza Benedict Spinoza "takes the idea of substance and works out its implications with logical consistency. If substance is that which needs nothing other than itself in order to exist or to be conceived, if God is *the* substance and everything else dependent on him, then, obviously, there can be no substance outside of God. Consequently, thought and extension cannot be separate substances, but must be attributes of the one single independent substance which is God. . . . the dualism of substances has disappeared, but the dualism of attributes remains. . . . Dualism gives way to monism, theism to pantheism, interactionism to parallelism" (Thilly and Wood, *A History of Philosophy*, pp. 320–21).

the rain drops . . . to have consciousness Confirmed source: Tyndall, "Science and Man," *Fragments of Science*, II, 356. Tyndall uses the image again in "Professor Virchow and Evolution," *Fragments of Science*, II, 408.

agnosticism See Huxley, *Hume*, p. 164, on the "one possible position," namely, agnosticism.

or carry out Berkeleys . . . mental phenomena Confirmed source: Huxley, *Hume*, pp. 169–71. Wilde continues to draw from Huxley on CN, 61.

* * *

[61]

no less than . . . to the body Confirmed source: Huxley, *Hume*, pp. 169–70.

whether there be not . . . on each incarnation Confirmed source: David Hume, "On the Immortality of the Soul," quoted in Huxley, *Hume*, pp. 172–73. Here Wilde joins together Huxley's discussion with Hume's speculation that there is a spiritual substance which nature uses as she employs matter, an idea which offers support for Wilde's belief in a racial unconscious, or Tribal Self. See CB, 119.

conception of the soul . . . or a ghostly This entry continues on CN, 63; confirmed source: Tyndall, "Professor Virchow and Evolution," *Fragments of Science*, II, 405, 408.

* * *

[62]

Δῖνος is the new God dynos, or whirling, held to be the opposite of *νους*, or rational mind, as the regulator of the universe. The idea may have been suggested or recalled from Aristophanes' *Clouds* (380) by Ruskin in "The Queen of the Air," *The Works of John Ruskin*, XIX, 326–37. Wilde's classical training amply acquainted him with Aristophanes; see his translation of the "Chorus of Cloud Maidens," CW, 816.

"Δι' ἐξεληλακως having driven out Zeus. Aristophanes, *Clouds*, 828.

Dagon Dagon was a fertility god worshiped in Canaan. He is mentioned in the Old Testament as one of the chief deities of the Philistines. See, e.g., Jud. 16:23–30, 1 Sam. 5:1–5, and 1 Chron. 10:8–10.

* * *

[63]

telegraph clerk sitting inside the brain Confirmed source: Tyndall, "The Belfast Address," *Fragments of Science*, II, p. 165, and Huxley, *Hume*, p. 78. The telegraph clerk also appears on p. 55 of this notebook.

they are not prepared . . . thought and extension Probable source: Tyndall, "Professor Virchow and Evolution," *Fragments of Science*, II, 408–9. On these pages Tyndall recognizes the problems in explaining consciousness from a materialist perspective and says that he does not object to the idea of imagining a soul: "I, for one, would not object to this exercise in ideality." He also remarks that "the mechanical philosopher, *as such*, will never place a state of consciousness and a group of molecules in the relation of mover and moved" (p. 408). Cf. Huxley, "On the Physical Basis of Life": "[T]he thoughts to which I am now giving utterance, and your thoughts regarding them, are the expression of molecular changes in that matter of life which is the source of our other vital phaenomena" (p. 138).

* * *

[65]

Gibbon admired Xenophon . . . drawn from him Confirmed source: Edward Gibbon, *Memoirs of My Life*, (London: Thomas Nelson & Sons, 1966), pp. 75–76.

D'Alembert proposed that . . . the vulgar overlook Confirmed source: Edward Gibbon, "Essai sur l'Étude de la Littérature" (1759) in *The Miscellaneous Works of Edward Gibbon* (London: John Murray, 1814), IV, 67–68; cf. the "Rise," CW, 1137.

D'Alembert Jean le Rond d'Alembert (1717–1783), French mathematician and philosopher, coeditor of the *Encyclopédie*, and a leading representative of the Enlightenment.

* * *

[67]

The Hindus . . . the Greek Sandracottus Possible source: Hegel, *The Philosophy of History*, pp. 229–32. Cf. the "Rise," CW, 1105.

Chandragupta Chandragupta (fl. 321–297 B.C.), founder of the Mauryan empire in India; described by Megasthenes, the ambassador of Seleucus I, who went on several embassies to India during the period 302–291 B.C.

Chinese History . . . revolt against authority Possible source: Hegel, *The Philosophy of History*, pp. 113–18, 157–60. Cf. the "Rise," CW, 1105.

* * *

[71]

These expressions . . . result of rationalism Possible source: Lecky, *History*, I, 294–98. It is more likely that Wilde is summarizing the views of many of the thinkers whose works he was reading.

* * *

[73]

and was a strong feature of the Italian Renaissance This phrase was added with a broad-nibbed pen.

State of nature• . . . his highest developement Possible source: Hegel, *The Philosophy of History*, pp. 88–91.

* * *

[75]

the captain . . . Roman Empire Confirmed source: Gibbon, *Memoirs*, p. 117. Gibbon, who resided in Hampshire, was a member of the Hampshire militia and so is referring to himself. The remark is also quoted in Charles-Augustin Sainte-Beuve, "Gibbon" (1853), *Profils Anglais*, préface d'André Turquet, 2nd edn. (London: J. M. Dent; New York: G. P. Putnam's Sons, 1907), p. 128.

Why histories are . . . view continually shifts Cf. the "Rise," CW, 1135.

* * *

[77]

αγωνισμα ς αει a declamation for always. Wilde has adapted Thucydides 1.22.4: "a posession forever rather than a declamation for the moment."

To recreate . . . task of the Historian Possible source: Tylor, *Primitive Culture*, p. 5.

After all individuals . . . [continued on p. 79] surroundings[,] which differ Probable source: Tylor, *Primitive Culture*, p. 11.

* * *

[79]

the fulness . . . combatants themselves Cf. the "Rise," CW, 1142.

* * *

[80]

Polybius remarks . . . does a sample Cf. the "Rise," CW, 1138.

* * *

[81]

as the anatomist . . . animals now extinct Possible source: Hippolyte A. Taine, "Introduction," *History of English Literature,* trans. H. Van Laum (New York: Hurst, 1873), p. 28. Taine says that a "clever naturalist can by process of reasoning reconstruct out of a few fragments almost the whole body. . . ." Cf. the "Rise," (CW, 1138), where Wilde reworks the language of the notebook passage.

* * *

[83]

the ordinary appeal . . . results will happen Possible source: Tylor, *Primitive Culture*, pp. 3–5.

* * *

[85]

the ancients regarded . . . preach a creed Possible source: Symonds, *Studies of the Greek Poets*, II, 380. See also Ruskin, "Queen of the Air," *The Works of John Ruskin*, XIX, 307–9, for a discussion of didacticism as a racial element. See also the "Rise," CW, 1115.

* * *

[86]

Columbus and Copernicus . . . degraded this planet Compare the "Rise," CW, 1148.

* * *

[87]

the study of Greek . . . and research Cf. the "Rise," CW, 1148.
except the blind . . . Greek in its origin Confirmed source: Symonds, *Stud-*

ies of the Greek Poets, II, 397. He quotes this comment from Sir Henry Maine's Rede Lecture of 1875.

* * *

[89]

R. the essentially plastic character of the Greek thinkers Probable source: Symonds, *Studies of the Greek Poets,* II, 376. There he quotes from Hegel, *The Philosophy of History*, p. 318: "The Greek Spirit is the plastic artist, forming the stone into a work of art." Cf. the "Rise," CW, 1137.

whose works are . . . versatility of Herodotus Cf. the "Rise," CW, 1137.

the modern mind . . . in creative power Confirmed source: Symonds, *Studies of the Greek Poets*, II, 364.

* * *

[91]

Michel Agnolo [*sic*] alone . . . painting and poetry See CB, 179.

* * *

[93]

the most remarkable . . . suit of leather Confirmed source: Thomas Carlyle, *Sartor Resartus* (1833–34), ed. C. F. Harrold (rpt. New York: Odyssey Press, 1937), pp. 208–9.

* * *

[95]

Archibald Alison wrote . . . of the Tories Confirmed source: James Anthony Froude, "Scientific Method Applied to History," *Short Studies on Great Subjects* (New York: Scribner's Sons, 1905), II, 468.

Archibald Alison Alison (1792–1867), author of *History of Europe During the French Revolution*, 10 vols. (1833–42), and other works.

Herodotus had a . . . to expose Imperialism Confirmed source: Froude, "Scientific Method," II, 465.

* * *

[97]

Not a single . . . is with <u>History</u> Confirmed source: Froude, "Scientific Method," II, 447–9. Cf. the "Rise," CW, 1113.

* * *

[99]

nec deus intersit . . . as of art Confirmed source: Froude, "Scientific Method," II, 449.

nec deus intersit nor let (a) god intervene

* * *

[100]

Greeks were Aryan . . . lands of Tibet Cf. the "Rise," CW, 1105.

* * *

[101]

Poets are the . . . poetry into prose Confirmed source: Froude, "Scientific Method," II, 450–60.

* * *

[103]

All religions are . . . statesman equally useful Confirmed source: Gibbon, *The Decline and Fall of the Roman Empire*, I, 34. Cf. the "Rise," CW, 1134n.

The Sybil . . . of her volumes Cf. the "Rise," CW, 1137.

* * *

[105]

The influence of . . . any particular belief Confirmed source: Lecky, *History*, I, 296.

But the force . . . remote from Law• Probable source: Lecky, *History*, I, 298–99. See also pp. 17–18. We believe this passage is fundamental for an understanding of Wilde's later writings. It suggests that the aesthetic sense makes both science and art possible: the "desire to bring all ones thoughts into harmony and mutual correspondance" is an internal need and not stimulated by actual patterns of natural order. This coincided neatly with Wilde's belief in the racial origins of the historic sense and the aesthetic sense, the critical and creative abilities, a belief he shared with his teachers J. P. Mahaffy and John Ruskin, as well as with many others. It also helps to explain Wilde's oft-asserted belief that creative and critical work are interrelated manifestations ultimately springing from the same faculty.

* * *

[107]

With Plato it . . . founded on psychology Cf. the "Rise," CW, 1121.
this system Aristotle . . . somewhat Hegelian way Cf. the "Rise," CW, 1128.
One of the . . . kinds of states Confirmed source: Aristotle, *Politics*, V, xii, 1316A, l. 1–6.

* * *

[108]

Herodotus' philosophy of . . . such as Bossuet's Cf. the "Rise," CW, 1115.
Bossuet Jacques Bénigne Bossuet (1627–1704), whose *Histoire Universelle* is perhaps the first attempt at a philosophy of history.
Polybius had really . . . a Roman Empire Cf. the "Rise," CW, 1132–33.
and from him . . . Dante and Mazzini Cf. the "Rise," CW, 1143.

* * *

[109]

The philosophy of . . . and the Persians Confirmed source: Ernest Renan, *The Life of Jesus* (New York: Modern Library, 1927), p. 101.

* * *

[111]

That no really . . . subjects drink wine Confirmed source: Montesquieu, "De l'esprit des lois," *Oeuvres Complètes* (New York: Macmillan, 1964), p. 539.
comme la grandeur . . . vie des empereurs as republican grandeur was fatal to republican government, imperial grandeur was fatal to the life of the emperors. Confirmed source: Montesquieu, "Sur les causes de la grandeur des Romaines et de leur décadence," *Oeuvres Complètes*, pp. 464–65.
on the free . . . athens and Rome Probable source: Montesquieu, "De l'esprit des lois," *Oeuvres Complètes*, 532–34, 536–37.
most important• we . . . on a battle Confirmed source: Montesquieu, "Sur les causes," *Oeuvres Complètes*, p. 472.

* * *

[113]

Herodotus: test of . . . her will 1 Confirmed source: Herodotus, *The History of the Persian Wars*, 1.4.2. See, e.g., the Beloe translation (New York: Harper, 1839), pp. 2–6, 27–29. Cf. the "Rise," CW, 1111.

If Helen had . . . her up. II. 118 Confirmed source: Herodotus, I, Bk. 2. Wilde's reference is to the beginning of the section on the Trojan War; the statement about Priam is at 2.120.2. See e.g., the Beloe translation, p. 258. Cf. the "Rise," CW, 1111.

φρενοβλαβης harmed in the mind, deranged, crazy

II. 134 Herodotus proves that Rhodopis the courtesan could not have built a pyramid, which was really built by Mycerinus; see Beloe, pp. 274–75. Cf. the "Rise," CW, 1111.

* * *

[115]

Ces details ne . . . hauteur d'une idee These details are not true to the letter; but they are true to a higher truth: they are more true than the naked truth in the sense that they are truth rendered expressive, verbal, elevated to the level of an idea. Confirmed source: Ernest Renan, "Vie de Jésus," *Oeuvres Complètes* (Paris: Calmann-Lévy, 1947), IV. 77.

So Aristotle says . . . truer than history Confirmed source: Aristotle, *Poetics*, 1451B, although Aristotle says "more philosophical and more serious." Cf. Froude, "Scientific Method," II, 462, on the speeches of Herodotus and Thucydides.

Notes

1. Richard Ellmann, for instance, in the published form of his recent lecture, *Oscar Wilde at Oxford* (Washington, D.C.: Library of Congress, 1984), does not mention the notebooks or many of the figures Wilde read and studied. In his recent, posthumously published biography, *Oscar Wilde* (London: Hamish Hamilton, 1987), Ellmann briefly mentions the Commonplace Book and, unfortunately, attributes to Wilde a few entries which the latter copied from others (pp. 40–41). It is regrettable that this magisterial biography appeared after our manuscript was already in proofs.

2. Walter Pater, *The Renaissance,* ed. Donald L. Hill (Berkeley: U of California P, 1980), p. 188.

3. Richard Ellmann, for instance, omits it from his collection of Wilde's criticism. However, a recent article in *Victorian Studies* (henceforth abbreviated as *VS*) focuses attention on one aspect of its importance. See Bruce Haley, "Wilde's 'Decadence' and the Positivist Tradition," *VS*, 28, no. 2 (Winter 1985), 215–29.

4. We will refer to the Commonplace Book in the text as CB and to the Notebook Kept at Oxford as CN, or College Notebook.

5. *16, Tite Street, Chelsea. Catalogue of the Library of Valuable Books . . . and numerous Effects. . . .* (London: Clarke, 1895), p. 8.

6. G. Thomas Tanselle, "Textual Scholarship," *Introduction to Scholarship in Modern Languages and Literatures,* ed. Joseph Gibaldi (New York: Modern Language Association, 1981), pp. 29–52; "The Center for Scholarly Editions: An Introductory Statement," *PMLA,* 92 (1977), 583–97.

7. Lewis R. Farnell, in *An Oxonian Looks Back* (London: Martin Hopkinson, 1934), recalls that the competitions in 1879 for open fellowships in classics at Jesus, Trinity, and Merton "drew together a large number of first-class men into intimacy, which expressed itself in the usual genial fashions of sympathy and mockery . . ." (pp. 70–71). Farnell remembers that Wilde was a member of this group.

8. These suggested dates—1874–76 for the College Notebook and the 1880s for the Commonplace Book—appear in the catalogue of the William Andrews Clark Memorial Library.

9. The essay was not published until 1909; the manuscript notebooks are in the William Andrews Clark Memorial Library. See H. Montgomery Hyde, *Oscar Wilde: A Biography*

(New York: Farrar, Straus, and Giroux, 1975), pp. 36–39. Hyde says that Wilde was still working on the essay in the early months of 1879.

10. There are two other notebooks which Wilde used in writing the "Rise." One, now housed at Yale University, begins with extensive notes in ink from Theodor Mommsen's *The History of Rome,* and also contains an early draft of Wilde's play *Vera, or the Nihilists* (1880), hastily scrawled in pencil on the pages remaining after the Mommsen notes but written from the back toward the front of the notebook. The primary placement of the Mommsen notes suggests that they were made before the *Vera* draft and therefore were written during Wilde's preparation of the essay. The second notebook, privately owned, contains notes on Greek and Roman history and philosophy. These notes are clearly related to topics in the "Rise," and there are passages which, corrected and rewritten, appear in the essay. The notebook is described in the Prescott Catalogue •411, quarto Autograph Notebook of 133 leaves (*The Prescott Collection, Printed Books and Manuscripts* [New York: Christie, Manson & Woods International, 1981], p. 171). We have decided not to edit these two notebooks because they add very little to what the Commonplace Book and the College Notebook show of Wilde's intellectual development or of the breadth of his reading. Two more notebooks housed at the Clark Library, both on philosophy, were not used in the writing of the "Rise." We have not edited them because they contain little of substance not already found in the Commonplace Book or the College Notebook. They are mainly composed of short, summary remarks on classical and modern philosophers, and we make reference to them occasionally.

11. The edition of Wilde's work we have used is *Complete Works of Oscar Wilde*, 2nd edn., ed. J. B. Foreman, intro. Vyvyan Holland (London: Collins, 1966); hereafter we will refer to it as CW. See, for example, the passages on Roman religion (CB, 73–75, and CW, 1146), and see also the comments on the origin of criticism (CB, 21, and CW, 1106).

12. John Charles Finzi, comp. *Oscar Wilde and His Literary Circle: A Catalog of Manuscripts and Letters in the William Andrews Clark Memorial Library* (Berkeley and Los Angeles: U of California P, 1957), #2427.

13. "To Lord Alfred Douglas" [Jan.–Mar. 1897], usually known as *De Profundis, The Letters of Oscar Wilde*, ed. Rupert Hart-Davis (London: Rupert Hart-Davis, 1962; New York: Harcourt, Brace & World, 1962), p. 466. Hereafter this edition will be referred to as *Letters*.

14. In adopting chronological order and dialectics, Wilde carefully differentiates his own position from Hegel's interpretation of history by stating, "not that the succession of time is *always* the succession of ideas, or that dialectics moves *ever* in the straight line in which Hegel conceives its advance" [emphasis added].

15. Thomas George Wilson, *Victorian Doctor: Being the Life of Sir William Wilde* (New York: L. B. Fisher, 1946), p. 133.

16. Hyde, *Oscar Wilde*, p. 4.

17. Wilson's *Victorian Doctor* and Horace Wyndham's *Speranza: A Biography of Lady Wilde* (London: T. V. Boardman, 1951) document the Wilde's extensive scientific and social interests and involvement. In *The Wildes of Merrion Square: The Family of Oscar Wilde* (London: Staples Press, 1953), Patrick Byrne reports that "William Hamilton was one of the supreme intellects of the nineteenth century, yet his name is hardly known today except as the author of a mathematical system. . . . He was one of Speranza's most intimate friends. . . . He had a long acquaintance with William Wilde, chiefly through a common membership of the Royal Irish Academy, and he had a high opinion of Wilde not only as an oculist but as a man of general culture" (pp. 48–50). For another, less flattering, picture of the parents, see Eric Lambert's *Mad With Much Heart: A Life of the Parents of Oscar Wilde* (London: Frederick Muller, 1967).

18. Hyde, *Oscar Wilde,* pp. 12–15, 20, 28–29. See also W. B. Stanford, *Ireland and the Classical Tradition* (Dublin, 1976; rpt. Dublin: Irish Academic Press, 1984), pp. 235–39. Letters Wilde wrote to his parents in 1875, the summer after he entered Oxford, provide further evidence of his interest in archeological science and historical studies. See *Letters,* pp. 4–10 passim.

19. *Twelve Lectures on Primitive Civilizations, and their Physical Conditions* (London: Longmans, Green, 1869), p. 222.

20. J. P. Mahaffy, *Prolegomena to Ancient History* (London: Longmans, Green, 1871), pp. 94–95. See also pp. 44–45 for Mahaffy's description of Müller's theory of Vedic and Hellenic mythological correspondence and its basis in the poetic naming of natural phenomena.

21. Mahaffy, *Prolegomena,* p.7.

22. Mahaffy, *Prolegomena,* p. 34. This conception of the historic sense may have informed Wilde's use of the idea in relation to Herodotus (CW, 1110).

23. Quoted in Hyde, *Oscar Wilde,* p. 14.

24. See Frank M. Turner, *The Greek Heritage in Victorian Britain* (New Haven: Yale U P, 1981), pp. 6–7. Turner's book offers an excellent overview of the various nineteenth-century interpretations of Hellenic history and culture. Students of Wilde will find his chapters on Greek religion, mythology, and the Platonic revival especially useful background for a proper appreciation of the notebooks.

25. Quoted in Hyde, *Oscar Wilde,* p. 17.

26. See F. Max Müller, *My Autobiography: A Fragment* (New York: Scribner's, 1901), pp. 130–42.

27. F. Max Müller, *Lectures on the Science of Language,* 2nd edn., rev. (New York: Scribner, 1868), I, 354; hereafter referred to in the text as SL.

28. Turner, *Greek Heritage,* p. 104.

29. Noel Annan, *Leslie Stephen* (Cambridge, Mass.: Harvard U P, 1952), pp. 150–61.

30. In *Sesame and Lilies,* his most popular book, Ruskin advises his readers on how to "deal with words rightly." It is necessary to study the etymology of words, which undergo "a certain change of sense and use on the lips of each nation" but still retain "a deep vital meaning, which all good scholars feel in employing them, even at this day." He goes on to say, "Read Max Müller's lectures thoroughly, to begin with; and, after that, never let a word escape you that looks suspicious." See John Ruskin, *Sesame and Lilies,* in *The Complete Works of John Ruskin,* ed. E. T. Cook and Alexander Wedderburn, Library Edition, 39 vols. (London: George Allen, 1903–12), XVIII, 68–69. Hereafter this edition will be referred to as *Works.* For other studies of Ruskin's theories of philology and mythology, as well as Müller's influence on him, see: Turner, *Greek Heritage,* p. 112; Elizabeth K. Helsinger, *Ruskin and the Art of the Beholder* (Cambridge, Mass.: Harvard U P, 1982), pp. 256ff.; and George Landow, *The Aesthetic and Cultural Theories of John Ruskin* (Princeton: Princeton U P, 1971), pp. 399–420.

31. John Ruskin, *The Queen of the Air,* in *Works,* XIX, 300–301. Hereafter referred to in the text as QA.

32. In different places in the text, Ruskin says that his ideas are not in conflict with Darwin's, that he accepts the protoplasmic theory as propounded by Huxley, and that he sees Tyndall's analysis of heat and of the blue coloring of the sky as confirmations of his own vitalist position (QA, 252–53, 354–59).

33. Gerald L. Geison, "The Protoplasmic Theory of Life and the Vitalist-Mechanist Debate," *Isis,* 60 (1969), 284.

34. Although there is much evidence that Ruskin accepted the theory of the inheritance of acquired characteristics, he does not explicitly state his assent in *The Queen of the Air.*

However, his insistence that "if the fathers have eaten sour grapes, the children's teeth are set on edge" (QA, 393)—a Biblical passage Wilde notes under "Heredity" (CB, 145), perhaps following Ruskin—and his description of the coherence of his theory with evolutionary science make his acceptance implicit. In the explanation of his vitalist interpretation of life and art, Ruskin notes: "The facts on which I am about to dwell are in nowise antagonistic to the theories which Mr. Darwin's unwearied and unerring investigations are every day rendering more probable" (QA, 358n). It is widely believed that Darwin rejected the Lamarckian theory of evolution by the inheritance of acquired characteristics in favor of the theory of natural selection. In fact, as the successive editions of *The Origin of Species* prove, Darwin came more and more to rely on the theory of inheritance as a supplemental explanation for evolutionary change.

35. It is possible that John Rosenburg ignored or was not aware of Ruskin's theory when he claimed that *The Queen of the Air* "does not conclude, but disintegrates under the weight of irrelevancies and delusions such as Ruskin's claim that his 'art-gift' belonged to him 'by birthright, and came by Athena's will.'" John D. Rosenberg, *The Darkening Glass: A Portrait of Ruskin's Genius* (New York: Columbia U P, 1961), p. 175.

36. The lectures were first published in *Works,* XXIII (1906), 179–279. The dates of the lectures appear on p. 182. This edition will be cited in the text as AM. In an interview published in the New York *Daily Tribune,* on 8 January 1882, Wilde recalled, "When I went to Oxford . . . Ruskin was then delivering a series of lectures on 'Florentine Art,' and I came under his influence." G. T. Atkinson, one of Wilde's contemporaries at Oxford, also recalled the lectures: "Ruskin in the winter of '74 was lecturing at the Museum and attracting a crowded audience of men and maidens. Wilde was always there." See E. H. Mikhail, ed., *Oscar Wilde: Interviews and Recollections,* 2 vols. (New York: Barnes and Noble, 1979), I, 39; 19.

37. In *Modern Painters,* II, Ruskin is painfully explicit about his racialist theory. He remarks that Englishmen have an instinctive aesthetic sense buried beneath their learned responses, but he carefully differentiates English instincts of beauty from the aesthetic responses of Negroes, since "no importance whatever is to be attached to the opinions of races who have never received any ideas of beauty whatsoever (these ideas being only received by minds under some certain degree of cultivation). . . ." See *The Complete Works of John Ruskin,* IV, 69–70. Hereafter referred in the text to as MP.

38. Lot 98 in the 1895 Auction Catalogue of Wilde's library contained "Ruskin's Modern Painters, vol. 2, and other books. . . ."

39. Rosenburg discusses these lectures in *The Darkening Glass,* pp. 176–78. The history of Ruskin's projected withdrawal and subsequent revision and republication of the second volume of *Modern Painters* may be found in the "Introduction" to this volume (*Works*, IV, xlviii–xlix). Very briefly, Ruskin became dissatisfied with the volume and planned to withdraw it from an edition planned for the 1870s, which was never published. However, it rose in his esteem as idealist aesthetics came under attack from modern materialist theories and he made it the basis of his 1877 lectures and later revised and reissued it in 1883.

40. Michael Levey, *The Case of Walter Pater* (London: Thames and Hudson, 1978), pp. 123–27; Richard Ellmann, *Golden Codgers: Biographical Speculations* (New York: Oxford U P, 1977), esp. pp. 50–51.

41. The history of the controversy may be found in Turner, *Greek Heritage*, pp. 322–68.

42. "Conclusion," *The Renaissance: Studies in Art and Poetry*, ed. Donald L. Hill (Berkeley: U of California P, 1980), pp. 188–89. Hereafter referred to in the text as TR.

43. Ruskin translates this sentence from the *Ethics*, Book X, Chapter 8: "And perfect

happiness is some sort of energy of Contemplation, for all the life of the gods is (therein) glad; and that of men, glad in the degree in which some likeness to the gods in this energy belongs to them. For none other of living creatures (but men only) can be happy, since in no way can they have any part in Contemplation.'' The word ''energy'' is a translation of the Greek *energeia*, a concept Ruskin finds less important than ''contemplation,'' the Greek *theoria* (MP II, *Works* IV, 7).

44. Ruskin, the Conrad of criticism, is everywhere suggestive but nowhere clear; first asserting that there are three functions of imagination, he quickly declares that the Associative imagination merely composes, and then claims that the Penetrative and Contemplative imaginations are ''closely connected'' and have a ''distinctive nature, that property which makes it imagination and not composition'' (MP II, *Works* IV, 228). That property can be understood as spirit, which is simultaneously cognitive and transformational.

45. Since giving his 1874 lectures on the Aesthetic and Mathematical schools (which did not appear in print until 1906), Ruskin had changed the word by which he denoted spiritual perception from ''aethesis'' to ''theoria.'' This change made no more than a nominal difference in the coherence of Ruskin's position, and it demonstrated further his opposition to the burgeoning aesthetic movement and to Pater.

46. Wilde's thirty-two-page holograph notebook on the *Ethics* of Aristotle is housed at the William Andrews Clark Memorial Library at U.C.L.A. Examples of his borrowings from Grant may be found in the opening summary of Book VI (which is drawn from *The Ethics of Aristotle Illustrated with Essays and Notes*, 2nd edn., rev., ed. Sir Alexander Grant, 2 vols. [London: Longmans, Green, 1866], II, 144) and in Wilde's notes for Chapter IV of Book VI on Τεχνε (Grant, II, 158n). Wilde focused particularly on issues having to do with the intellectual, moral, and artistic virtues, or *αρεται*. He also noted ideas having to do with the imagination, production, and perception of art. In 1886 Wilde published in the *Pall Mall Gazette* a list of ''Books not to read at all'' which included ''Grant's Aristotle, . . . all argumentative books, and all books that try to prove anything.'' Reprinted as ''To Read, or Not to Read,'' *The Artist as Critic: Critical Writings of Oscar Wilde*, ed. Richard Ellmann (New York: Vintage Books, 1968), p. 27.

47. Farnell, *An Oxonian Looks Back*, p. 43. In the following four pages Farnell describes an undergraduate's enthusiastic immersion in Oxford Hegelianism.

48. J. E. Chamberlin, *Ripe Was the Drowsy Hour: The Age of Oscar Wilde* (New York: The Seabury Press, 1977), p. 58.

49. William Wallace, ''Prolegomena,'' *The Logic of Hegel* (Oxford: Clarendon Press, 1874), pp. cxxvii–viii. Hereafter referred to in the text as ''Prolegomena.''

50. Benjamin Jowett, ''Introduction'' to Plato's ''Sophist,'' *The Dialogues of Plato,* 2nd edn., trans. with analyses and introductions by B. Jowett, 5 vols. (Oxford: Clarendon Press, 1875), IV, 373. Hereafter referred to in the text as DP.

51. In his introduction to ''The Symposium'' Jowett added to Wallace's synthetic analysis of Darwin and Hegel, and Wilde noted (CB, 204) Jowett's ideas of the development of *ερως (eros)* from sexual urge to its sublimation in mystical and spiritual thought. On conflicting nineteenth-century interpretations of ερως in ''The Symposium,'' see Turner, *Greek Heritage,* pp. 397, 424–26.

52. Although Lyell apparently coined the phrase in 1851 for scientific usage, Jowett's use of ''missing link'' in this passage is cited by the OED as the first usage of this evolutionary term in a philosophical sense.

53. There is no conclusive external evidence that Wilde knew Hegel's *Aesthetik (Philosophy of Fine Arts)* firsthand, but it remains a possibility. The introduction to it was not translated into English until 1886. However, Bernard Bosanquet, the translator, lectured at Oxford during Wilde's residence, and there was a French translation by Charles Benard (5

vols., Paris, 1840–52) which Wilde could easily have read. A summary of Hegel's philosophy of art appears in sections 556–63 of the *Encyclopaedia of the Philosophical Sciences,* translated by William Wallace as *Hegel's Philosophy of Mind* (Oxford: Clarendon Press, 1971), pp. 293–97. There was a partial English translation of the *Aesthetik* by W. M. Bryant (New York, 1879). Furthermore, as we will later show, Wilde's published criticism bears the unmistakable stamp of Hegel in its special use of terms and the general use of arguments.

54. His transcription (CB, 171) of Pater's line "For us necessity is not as of old an image without us" came from either the 1867 or 1873 versions; thereafter the passage was changed. For a record of the changes, see the textual notes in Donald Hill's edition of *The Renaissance: Studies in Art and Poetry: The 1893 Text* (Berkeley: U of California P, 1980), p. 271. The similarity between Pater's language (which appeared only in the *Westminster Review,* 87 [January–April 1867], 43 [American edition]) and Wilde's entry (in CB, 153) is persuasive but not conclusive, providing evidence that Wilde based his entries on the 1867 version.

55. For an account of Symonds at Oxford, see Phyllis Grosskurth, *The Woeful Victorian: A Biography of John Addington Symonds* (New York: Holt, Rinehart and Winston, 1964), pp. 42–68. The diary entries are reported on p. 53.

56. *The Letters of John Addington Symonds,* ed. Herbert M. Schueller and Robert L. Peters, 3 vols. (Detroit: Wayne State U P, 1967), I, 401, 783. Hereafter referred to in the text as JASL.

57. See *Letters,* p. 25 and note.

58. As Phyllis Grosskurth points out, evolutionary theory also offered Symonds a scientific explanation, expressed in the last essay of *Studies,* for his own homosexual tendencies; see *The Woeful Victorian,* p. 163.

59. *The Essays of Oscar Wilde* (New York: H. S. Nichols, 1916), p. 426.

60. Henry Paolucci, ed. and trans., *Hegel: On the Arts* (New York: Frederick Ungar, 1979), p. xvii. Hereafter cited in the text as "A."

61. "Everything," said Hegel, "that is genuinely true, in mind as in nature, is inherently concrete, having both subjectivity and particularity in itself, as well as universality" (A, 8).

62. That Wilde understood this concept is clear from his remark in CB, 206: "The cognition of sensuous objects need not itself be sensuous."

63. Hegel, *Aesthetics: Lectures on Fine Art,* trans. T. M. Knox, 2 vols. (Oxford: Oxford U P, 1975), II, 1158. Hereafter referred to in the text as LFA.

64. Wilde may have developed this position from his reading of Symonds' volumes on the Greek poets, which, as Frank Turner notes, "provide a striking example of Hellenism breaking beyond the bounds of respectable domestication" (Turner, *Greek Heritage,* p. 114). Wilde anticipated the reinterpretation of Hellenism developed by Jane Harrison and the Cambridge anthropologists (ibid., p. 127).

65. In his published writings, letters, and in the notebooks, Wilde consistently disparaged Mill and other utilitarians. (See, e.g., CB, 129 on Mill and Bentham and CB, 15, 39, and 121 on utilitarian consequentialist ethical theory).

66. Here Spencer's position about the relationship of language to reality converged with conclusions Jowett derived only through Hegel's *Logic.* For both men the inadequacies of linguistic categories to describe the world offered an escape from the dominion of abstract ideas and fixed categories.

67. This theory permeated Spencer's works; in relation to mind, see *The Principles of Psychology,* 3rd edn., 2 vols. (London, 1872 and 1880; rpt. New York: Appleton, 1905), I, 470–71. Spencer accounts for reason, intelligence, and creativity as follows: "[T]he

human brain is an organized register of infinitely-numerous experiences received during the evolution of life, or rather, during the evolution of that series of organisms through which the human organism has been reached. The effects of the most uniform and frequent of these experiences have been successively bequeathed, principal and interest; and have slowly amounted to that high intelligence which lies latent in the brain of the infant—which the infant in after life exercises and perhaps strengthens or further complicates—and which, with minute additions, it bequeaths to future generations. And thus it happens that the European inherits from twenty to thirty cubic inches more brain than the Papuan. Thus it happens that faculties, as of music, which scarcely exist in some inferior human races, become congenital in superior ones. Thus it happens that out of savages unable to count up to the number of their fingers, and speaking a language containing only nouns and verbs, arise at length our Newtons and Shakespeares." Elsewhere Spencer accounts for altruism and the moral sentiments as products of "the evolution of mind by the accumulated and inherited effects of experiences" (II, 618).

68. Maurice Mandelbaum provides a useful description of the Lamarckian evolutionary theory Wilde adopted from Spencer:

> In a Lamarckian theory . . . the need for adaptation is the basis on which new structures gradually arise, and the lack of a need for a particular structure eventually causes it to disappear. Thus adaptation is a positive factor, being the tendency of an organism, impelled by its inner needs, to grow into closer relationship with its environment. Thus, on Lamarck's theory, there is a tendency on the part of every type of animal to make itself into that which, given a particular environment, it will eventually become. This Lamarckian view of evolution was characteristic of Herbert Spencer's biology, and of his psychology as well. . . .
> . . . Spencer, looking at the process of evolution as a whole, rather than considering what could be observed with respect to individual organisms and their offspring, was willing to allow far wider scope to the inherited effects of experience than Darwin had done. This difference is particularly evident with respect to Spencer's views regarding the evolution of mind. . . . For Spencer it was even possible that, within the scope of modern history, new capabilities had developed through habit, and that these capabilities had—within that relatively short time—become inherited traits." *History, Man, and Reason: A Study in Nineteenth-Century Thought* (Baltimore: Johns Hopkins U P, 1971), pp. 230–32.

69. Herbert Spencer, *The Study of Sociology* (New York and London, 1873; rpt. New York: D. Appleton, 1884), p. 338.

70. Mikhail, *Oscar Wilde,* I, 107.

71. In "The Philosophy of the Pure Sciences" William Kingdon Clifford quotes extensively Spencer's theory in *The Principles of Psychology,* remarking, "This doctrine of Mr. Spencer's is what I believe to be really the truth about the matter. . . ." *Lectures and Essays,* ed. Leslie Stephen and Frederick Pollock, 2 vols. (London: Macmillan, 1879), I, 279–81. Hereafter this edition will be referred to in the text as L & E. Wilde read the essay and made many notes from it in his Commonplace Book; see pp. 119, 125, 137, 139, 187, and 207.

72. Clifford was not the only idealist philosopher of science Wilde read. His manuscript Notebook on Philosophy at the William Andrews Clark Memorial Library contains entries based on the works of William Whewell and Sir William Hamilton. A few comments show Wilde's interest in the contemporary philosophical problems which he takes up in the College Notebook and the Commonplace Book. For instance, on mind and matter he remarks, "If intelligence in man is only product of matter, then in universe the phenomena of intelligence by design are only the products of a brute necessity" (Notebook on Philosophy, p. 10).

73. Maurice Mandelbaum, in *History, Man, and Reason,* pp. 10–20, esp. pp. 17–19,

describes the general reinterpretation of positivism during and after the time at which Clifford wrote.

74. Pollock, "Introduction," *Lectures and Essays,* I, 39.

75. The similarity of Clifford's and Wilde's positions to Spencer's may be seen in the following passage from Hugh Elliot, *Herbert Spencer* (New York: Henry Holt, 1917), pp. 277–78, on Spencer's attempt to reconcile empiricism and transcendentalism: "Spencer held that Locke was right in saying that all intelligence is derived from experience, but it is the experience, not of the individual, but of the race. It was equally true, as Kant said, that intelligence is largely preordained as regards the individual; but that prearranged disposition is not transcendental in character—it is the accumulated result of the experience of former generations. According to Spencer, therefore, intelligence is *à priori* as regards the individual and *à posteriori* as regards the race."

76. Thomas Henry Huxley, *Hume* (New York: Harper & Bros., 1879), p. 78.

77. For other entries on the materialist and idealist explanations of mind and matter, see CB, 89, 93–95, 105, 147, and 149.

78. Richard Ellmann, "The Critic as Artist as Wilde," *The Artist as Critic,* pp. xiii–xvii.

79. Ibid., p. xvi.

80. Walter Pater, "Winckelmann," *The Westminster Review* [American edn.] 87 (January–April 1867), 47. Hereafter referred to in the text as W. See CB, 206.

81. Turner, *Greek Heritage,* pp. 16–17.

82. Turner, *Greek Heritage,* p. 61. The earliest example of Wilde's evolutionary and racialist interpretation may be found in an essay he wrote at Oxford in 1877 entitled, "Hellenism." (Unpublished in his lifetime, it first appeared in a limited edition printed for subscribers by the Tragara Press, Edinburgh, 1979.) In the essay he describes the origins of the Hellenes and characterizes them as a "community of blood and language, and religious feelings and festivals and general character." Hellenism, he continues, is "a one-sided enthusiasm for ideas extending over a multitude of peoples" (pp. 7–8). The Hellenes can be defined only in terms of "common feelings and sentiment; they never produced any political union" (p. 7). In fact, the "intense individuality" of the Greeks developed because of the geographical and political isolation of the city-states. This brief essay is important because Wilde identifies Hellenic culture with individuality and not with constitutional order. For Wilde culture is perfected when individuals are free to seek self-realization. This attitude is consistent with his opposition to the Arnoldian assumption that culture must be developed through civil order and that individualism leads only to Philistinism and chaos.

83. See, e.g., Wilde's entry on "Thought," CB, 2.

84. Henry Thomas Buckle, *History of Civilization in England,* 2 vols. (New York: D. Appleton, 1880), II, 418–19. Hereafter referred to in the text as HCE.

85. Robert Flint, *The Philosophy of History in France and Germany* (Edinburgh and London: Blackwood and Sons, 1874), p. 8.

86. Alice I. Perry Wood ("Oscar Wilde as a Critic," *North American Review,* 202 [July–December 1915], 899–909), calls attention to ways in which Wilde's criticism anticipates postimpressionism and cubism in art and symbolism in literature: "He is touched by the modern spirit in many ways, he suggests in his work many phases of recent thought upon art and literature" (909). Recently the idea has been taken up by several critics. Hilda Schiff ("Nature and Art in Oscar Wilde's 'The Decay of Lying,' " *Essays and Studies, Volume 18,* ed. Sybil Rosenfeld [London: John Murray, 1965], pp. 83–102) claims that Wilde's "contribution to, or at least anticipation of, the swing in modern criticism towards

the conception and evaluation of literature as organized images or symbols has been grossly undervalued'' (p. 83). In ''The Critic as Artist as Wilde,'' Richard Ellmann calls Wilde an ''ancestral Northrop Frye or Roland Barthes'' and claims he ''laid the basis for many critical positions which are still debated in much the same terms, and which we like to attribute to more ponderous names'' (*The Artist as Critic,* p. x). R. J. Green (''Oscar Wilde's *Intentions:* An Early Modernist Manifesto,'' *British Journal of Aesthetics,* 13 [1973], 397–404) suggests that *Intentions* is an ''early example of the beginnings of a literary and artistic revolt that was to reach its climax, in England, between 1910 and 1915 with the work of Lewis, Pound, T. E. Hulme, Epstein, Gaudier-Brzeska and Ford's *Good Soldier''* (397). Herbert Sussman (''Criticism as Art: Form in Oscar Wilde's Critical Writings,'' *Studies in Philology,* 70 [1973], 108–22) says that ''the ceaseless movement of the Wildean dialogue . . . dramatizes Wilde's modernist view that intellectual activity proceeds not by the rules of logic but is shaped by deeper psychic forces'' (114). Bruce Bashford's articles (''Oscar Wilde, His Criticism and His Critics,'' *ELT,* 20, no. 4 [1977], 181–87, enlarged as ''Oscar Wilde and Subjectivist Criticism,'' *ELT,* 21, no. 4 [1978], 218–34) develop an explanation of Wilde's critical theory which recognizes its complex and systematic nature. However, Bashford identifies formalism—and its problematic relationships with creative and critical acts—as the heart of Wilde's interest. In this sense he sees Wilde as a precursor to ''some of the most exciting work in contemporary criticism [which] is now focusing on the reader's competence, response, and subjectivity'' (*ELT* 20, no. 4, 186). Regenia Gagnier (*Idylls of the Marketplace: Oscar Wilde and the Victorian Public* [Stanford: Stanford U P, 1986]) suggests that Wilde's language, ''which deconstructed bourgeois categories of thought,'' and his essays anticipate Bakhtin's idea of the ''dialogical'' (pp. 31–32).

87. Wendell V. Harris, ''Arnold, Pater, Wilde, and the Object as in Themselves They See It,'' *Studies in English Literature, 1500–1900,* II (Autumn 1971), 733–47; Ellmann, ''The Critic as Artist as Wilde,'' pp. xi–xii.

88. Critics have modified the extreme solipsism and relativism of this position by saying Wilde and Pater were inconsistent. See Patricia Clements, *Baudelaire & the English Tradition* (Princeton: Princeton U P, 1985), p. 160; see also Harris, ''Arnold, Pater, Wilde,'' 747.

89. Walter Pater, ''Coleridge,'' in *Appreciations,* in *English Prose of the Victorian Era,* ed. C. F. Harrold and W. D. Templeman (New York: Oxford U P, 1938), p. 1438. Hereafter referred to in the text as H & T.

90. Philip Appleman, ''Darwin and Pater's Critical Dilemma,'' *Darwin: A Norton Critical Edition,* ed. Philip Appleman (New York: Norton, 1970), p. 616.

91. Ibid., p. 616. Peter Allan Dale (*The Victorian Critic and the Idea of History: Carlyle, Arnold, Pater* [Cambridge, Mass.: Harvard U P, 1977], pp. 206–45) and F. C. McGrath (*The Sensible Spirit: Walter Pater and the Modernist Paradigm* [Tampa: U of So. Florida P, 1986], 122–23) discuss Pater's debts to and differences with Hegel. Neither believes that Pater uses Hegel to resolve the contradiction which Appleman describes.

92. Anthony Ward, *Walter Pater: The Idea in Nature* (London: MacGibbon and Kee, 1966), pp. 76–77.

93. See also our previous discussion of Wilde's entry on the function of analogy in physical science (CN, 105). When Wilde was in the United States in 1882, his remarks to a reporter show that he regarded the relation of science and poetry in much the same way as he did in the notebooks: ''Poets, you know, are always ahead of science; all the great discoveries of science have been stated before in poetry. So far as science comes in contact with our school [i.e., the aesthetic school], we love its practical side; but we think it absurd

to seek to make the material include the spiritual, to make the body mean the soul, to say that one emotion is only a secretion of sugar, and another nothing but a contraction of the spine.'' Quoted in *Oscar Wilde: Interviews and Recollections,* ed. E. H. Mikhail, I, 45.

94. Ellmann, ''The Critic as Artist as Wilde,'' pp. xiii, xvii–xviii. In *Oscar Wilde at Oxford* (1984), Ellmann continues to portray Wilde as a writer who based his work in unresolved conflicts and attitudes, not in a philosophical synthesis which could reconcile conflicts. He says, ''Wilde writes his works out of a debate between doctrines rather than out of doctrine'' (p. 27).

95. ''The English Renaissance of Art,'' *The Essays of Oscar Wilde,* (New York: H. S. Nichols, 1916), pp. 443–81, hereafter referred to in the text as ER. ''The Decorative Arts'' and ''The House Beautiful'' appear in Kevin O'Brien, *Oscar Wilde in Canada: An Apostle for the Arts* (Toronto: Personal Library, 1982), pp. 147–81. Critics, for the most part, have limited their substantial comments on the lectures to remarks about Wilde's indebtedness to Ruskin and his disciple, William Morris, for specific ideas. Wilde self-consciously remarked, in his review of Pater's *Appreciations* (1890), that Pater had ''escaped disciples.'' See Ellmann, *The Artist as Critic,* p. 234. Rodney Shewan (*Oscar Wilde, Art and Egotism* [New York: Barnes & Noble, 1977], pp. 20–22) does recognize the continuity of ''The English Renaissance in Art'' with the rest of Wilde's critical writings, but he does not see that Wilde's intellectual synthesis informs the analytic categories, the historical perspective, and so the structure and development of the lecture.

96. O'Brien, *Oscar Wilde in Canada,* pp. 26–27, 150, 190–93. While we agree with O'Brien that Wilde's ideas in the lectures and in *Intentions* are similar, we disagree with his account of the substance of that continuity as ''the element in the Aesthetic Movement that became dominant in the 1890s—the 'art for art's sake' theory of isolationists who valued beauty and art as the only noble goals in life and who rejected society's demands on the artist'' (pp. 34–35).

97. Hyde, *Oscar Wilde,* p. 46.

98. Wilde does not mention Pater's name anywhere in the lectures; his references to and disagreements with Pater's ideas are politely covert. In contrast, he openly and approvingly quotes Ruskin. Wilde's adoption of Ruskin's ideas over Pater's began at Oxford and was publicly expressed in Wilde's review of ''The Grosvenor Gallery, 1877,'' which he ends by referring to ''that revival of culture and love of beauty which in great part owes its birth to Mr. Ruskin, and which Mr. Swinburne, and Mr. Pater, and Mr. Symonds, and Mr. Morris, and many others, are fostering and keeping alive, each in his own peculiar fashion.'' *The Essays of Oscar Wilde,* p. 231.

99. The abridgment of the lectures is described in O'Brien's *Oscar Wilde in Canada,* pp. 147–51. Wilde began the lecture tour by writing a long (and, as he found out, boring) lecture which relied heavily on historical analysis and philosophical explanation. He presented himself as a learned man, only to find that the public wanted an entertainer. To satisfy his audiences, Wilde cut much of his ''learning'' from the lectures, but comparison of the first and later versions of ''The English Renaissance of Art'' demonstrates Wilde's fidelity to the theoretical coherence which his synthesis provided. Since the longer original version embodied the fullest exposition of Wilde's thoughts, we will use it as our text. The later version we have consulted is ''Lecture on the English Renaissance'' in *The Works of Oscar Wilde,* Sunflower edn., 13 vols. (New York: Lamb Publishing Co., 1909), *Poems in Prose,* pp. 103–17. We will refer to this edition as ER2.

100. See E. T. Cook, *The Life of Ruskin,* 2 vols. (New York: Macmillan, 1911), I, 339.

101. Billie Andrew Inman (*Walter Pater's Reading: A Bibliography of His Library Borrowings and Literary References, 1858–1873* [New York: Garland, 1981], p. 139) suggests

Plato as the source for "the tyranny of the senses" and Hegel's *Aesthetik*, Pt. 1, Introduction, 3, A, 2 as the source for the idea of escaping from the tyranny of the senses. As Inman shows, Pater would also have been aware of Hegel's dependence on *theoria* (p. 139).

102. This passage is based almost verbatim on the entry "Plastic Art," CB, 173.

103. Wilde suggested the possibility of this escape and renewal through art in his poem "Theoretikos" (CW, 31); later he had Lord Henry Wotton express the same idea aphoristically and paradoxically in *The Picture of Dorian Gray:* "[T]hat is one of the great secrets of life—to cure the soul by means of the senses, and the senses by means of the soul" (CW, 140–41).

104. Wilde borrows the marriage of Faust and Helen from Pater's "Winckelmann" (W, 40). Pater used Euphorion to symbolize Goethe's symbolic role as the progenitor of nineteenth-century art, but Wilde changed its application and had Euphorion symbolize the English renaissance.

105. See CN, 23 for Wilde's entry on science and poetry, where he uses the phrase "definite conception." The notebook entry is a possible source for the lecture.

106. Wilde later revised this comment, perhaps because it offended his audience. In "The Decorative Arts" he says, "I do not regard the commercial spirit of the present age as being opposed to the development of art" (O'Brien, *Oscar Wilde in Canada,* p. 156). But in the same lecture and in "The House Beautiful" he continues to criticize "materialism" and the "greed for gain" because they destroy the environment and oppress the spirit (ibid., pp. 164, 180).

107. In his first lecture Wilde refers simply to "Northern races," but in the later version he is more specific: "Teutonic and Saxon races" (ER2, 115).

108. Wilde borrows Goethe's remark about Winckelmann from Pater's essay (W, 38), but he adapts it to his own ideas, changing individual benefit to the common good.

109. Wilde quotes from Plato's *Republic*; he refers to the same "scheme of the noblest education" at CB, 185. In "The Decorative Arts" Wilde reformulates the idea in his own language (O'Brien, *Oscar Wilde in Canada,* p. 163).

110. This contradiction appears throughout "Winckelmann" in Pater's use of the term "spirit." He employs the Hegelian vocabulary of art history but means by "spirit" only the common characteristics embodied by the art of a period or the collective psychological and emotional needs satisfied by art.

111. William K. Wimsatt, Jr., and Cleanth Brooks, *Literary Criticism: A Short History* (1957; rpt. New York: Knopf, 1964), p. 485.

112. René Wellek, *A History of Modern Criticism: 1750–1950,* 5 vols. (New Haven: Yale U P, 1965), IV: *The Later Nineteenth Century,* p. 409. Hereafter referred to in the text as HMC IV.

113. Jacques Maritain also noted a "shoddy Hegelianism" in Wilde's theory. See "The Frontiers of Poetry" in *Art and Scholasticism,* trans. J. F. Scanlan (New York: Charles Scribner's Sons, 1942), p. 133. More recent critics who note in passing a Hegelian influence in Wilde's criticism are J. E. Chamberlin, *Ripe Was the Drowsy Hour,* pp. 22, 58, and Rodney Shewan, *Oscar Wilde, Art and Egotism,* pp. 19, 102.

114. Wilde used his historicist approach to defend a specific cause, namely, the movement in dramatic production (encouraged by the appearance in 1881 of the Saxe-Meiningen company and led by Henry Irving's Lyceum group) which emphasized the unity of artistic illusion through historically accurate settings and costumes. Bulwer-Lytton's criticism of the archeological accuracy of costume and setting in "Mary Anderson's Juliet" (*The Nineteenth Century,* December 1884) spurred Wilde's rebuttal.

115. A reaffirmation of this belief appears in the last paragraph of the essay where Wilde

writes: "Of the value of beautiful costume in creating an artistic temperament in the audience, and producing that joy in beauty for beauty's sake, without which the great masterpieces of art can never be understood, I will not here speak . . ." (CW, 1078). Here Wilde emphasizes that the "joy in beauty for beauty's sake" has the purpose of clarifying the truth of a work of art; that is, without this aesthetic sense, a masterpiece cannot be understood. So, later in the paragraph, Wilde insists: "we shall require of [future critics], indeed, that they cultivate a sense of beauty" (CW, 1078).

116. Charles Taylor, *Hegel* (Cambridge: Cambridge U P, 1975), p. 468.

117. "Shakespeare and Stage Costume," *The Nineteenth Century,* 17, no. 99 (May 1885), 818; (CW, 1078).

118. Oscar Wilde, *Letters,* pp. 294–95. Had he done the same with the English editions, scholars might have been more likely to see the relation between Wilde's aesthetic and political theories.

119. Hilda Schiff, "Nature and Art in Oscar Wilde's 'The Decay of Lying,' " *Essays and Studies, Volume 18,* p. 101. Bruce Bashford, *ELT,* 21, no. 4 (1978), 219, 221.

120. *More Letters of Oscar Wilde,* ed. Rupert Hart-Davis (New York: Vanguard Press, 1985), p. 80.

121. *Letters,* p. 236.

122. Shewan, *Art and Egotism,* p. 97. Shewan has both understood and misunderstood Wilde's rhetorical and intellectual tactics in "The Decay of Lying." In his belief comments he recognizes the continuity of doctrine between this essay and "The English Renaissance of Art" as well as the importance (if not the full significance) of dialectic, exaggeration, and the systematic inversion of meaning in Wilde's practice. However, Shewan does not see how Wilde's performing persona, argument, and rhetorical tactics are creative expressions of a systematic critical position.

123. This important observation did not appear in the first publication of the dialogue in *The Nineteenth Century,* 25 (January 1889), 35–56. It was added, along with considerable other material, to the revised version in *Intentions.*

124. As usual, Wilde's allusions are wittily significant. It was Esau who sold his birthright to his brother Jacob, one of the great Biblical liars, for a mess of pottage. Jacob knew the value of costumes: He dressed himself in animal skins and lied to convince the blind Isaac that he, Jacob, was the hairy Esau who should receive the paternal blessing. See Gen. 25:29–34 and 27:15–29.

125. For a discussion of the links between culture and race in the thought of Arnold and Wilde, see Frederic Faverty, *Matthew Arnold the Ethnologist* (Evanston, Il.: Northwestern U P, 1951); Lionel Trilling, *Matthew Arnold* (New York 1939, 1949; rpt. New York: Meridian Books, 1955); and the section in this volume entitled "Anarchy and Culture: The Evolutionary Turn of Wilde's Cultural Criticism."

126. Wilde's creative rhetorical tactics even extend to the naming of the speakers after his young sons. If he played a private joke by using his sons' names, there is also an openly resonant pun on "Vivian" which suggests "vivo," Latin for life or living well. Or perhaps he chose that name for his child because of its meaning. Vivian, appropriately, preaches the synthesis of nature and art as realized in life.

127. See, for example, Richard Ellmann, "The Critic as Artist as Wilde," *The Artist as Critic,* p. xxi, and Rodney Shewan, *Art and Egotism,* pp. 70–78.

128. Quoted in Martin Tucker, ed., *The Critical Temper: A Survey of Modern Criticism on English and American Literature from the Beginning to the Twentieth Century,* 3 vols. (New York: Ungar, 1969), II, 504.

129. Wilde himself concludes "that curious love of green . . . in individuals is always

the sign of a subtle artistic temperament, and in nations is said to denote a laxity, if not a decadence of morals'' (CW, 996).

130. For another example see ''The Portrait of Mr. W. H.'' (CW, 1150).

131. In ''Pen, Pencil and Poison'' he briefly foreshadows this broader social perspective of criticism when he remarks, ''Crime in England is rarely the result of sin. It is nearly always the result of starvation'' (CW, 1006).

132. Wilde echoes Herbert Spencer's conclusions about retrogression and progression in the evolution of various cultures. See Herbert Spencer, *The Evolution of Society: Selections from Herbert Spencer's Principles of Sociology,* ed. Robert L. Carneiro (Chicago: U of Chicago P, 1967), p. xlii.

133. See CB, 14; compare Wilde's position in ''The Decay,'' as applied to the development of art, which holds that the social organism, like untutored nature, is dominated by an imitative instinct, and that only the presence of mind could make evolution progressive. This somewhat conservative position summarizes the consensus of the first generation of English thinkers, who used the theory of evolution to explain social development. See J. B. Burrow, *Evolution & Society: A Study in Victorian Social Theory* (Cambridge: Cambridge U P, 1966), pp. 93–100.

134. This spirit is made self-conscious through the effects of living in a beautiful environment (CW, 1049; see also CB, 40, 185). Rodney Shewan has recognized that health is a central concept of Wilde's critical theory: ''A healthy state depends on the health of individuals, which depends in turn on full self-realisation'' (*Art and Egotism,* p. 99).

135. In the poem life in England, ''this vile traffic-house,'' like the essay's ''sphere of practical and external life'' (CW, 1040), poisons the soul. To escape spiritual pollution the soul must realize both its dilemma and its heritage of racial experience. Even if ''rude people rage with ignorant cries / Against an heritage of centuries,'' the speaker of the poem, like Wilde's elect, the self-conscious critics and artists, can still learn to know both self and, implicitly, the inherited collected life of the race. The speaker commands the soul to leave ordinary life and contemplate ''dreams of Art / And loftiest culture'' (CW, 716).

136. This is perhaps Wilde's answer to the rhetorical question he asks in his notebook: ''[I]s the voice of one crying in the wilderness merely the result of the molecular action of locusts and wild honey?'' (CN, 56).

137. In his notebooks Wilde includes the love of beauty in his metaphysics: ''[T]he Love of the beautiful,'' he says in the Commonplace Book, is ''the Love which Plato regarded as the essential concomitant of philosophy'' (CB, 69). In an entry on the three stages of Platonic ideas (logical, ontological, and poetical) he notes that ''from the love of the beautiful object we rise to the ideal . . .'' (CB, 193).

138. Cf. CB, 69: ''Survival of Fittest in thought.''

139. Charles Darwin, *The Descent of Man and Selection in Relation to Sex,* 2nd edn. (London, 1874; rpt. New York: A. L. Burt, n.d.), p. 706. Hereafter referred to as DM. Lot number sixty-three in the 1895 auction catalogue of Wilde's library is ''Darwin's Origin of Species and Descent of Man, 4 vols. illus.''

140. Darwin's use of the term ''man'' and the masculine case pronoun ''he'' in his description of sexual selection implicitly appropriates the power of choice for males and away from females. His language shows that man is superior to animals in sexual selection not only because ''he'' rather than ''she'' chooses but because ''he highly values mental charms and virtues. On the other hand, he is strongly attracted by mere wealth or rank'' (DM, 706). Darwin's description suggests the standard Victorian social practice: Men select from a number of potential mates the female they wish to propose to; female choice is limited to acceptance or rejection of a proposal.

141. The relationship of aesthetics to sexual selection in Wilde's analogy is consistent with his acceptance of Spencerian evolutionary theory and Hegelian idealism, as well as with Darwin's notion that sexual selection depended upon and even influenced the development in humans of a sense of taste and an appreciation for the beautiful.

142. This section is a revised version of the article "Anarchy and Culture: The Evolutionary Turn of Cultural Criticism in the Work of Oscar Wilde," which originally appeared in *Texas Studies in Literature and Language,* 20, No. 2 (Summer 1978), 199–215.

143. Raymond Williams, *Culture and Society, 1780–1950* (1958; rpt. New York: Harper & Row, 1966), p. 161. Hereafter referred to in the text as C & S.

144. For a bibliographical survey of Arnold's influence on Wilde's criticism, see Wendell V. Harris, "The Critics," *Victorian Prose: A Guide to Research,* ed. D. J. DeLaura (New York: Modern Language Association, 1973), pp. 461–63.

145. Northrop Frye, "The Problem of Spiritual Authority in the Nineteenth Century," *Backgrounds to Victorian Literature,* ed. Richard A. Levine (San Francisco: Chandler, 1967), p. 134.

146. For Arnold's reliance upon the "sciences" of ethnology and philology, see Frederic Faverty's lucid and scholarly study *Matthew Arnold the Ethnologist.* See also F. J. W. Harding, *Matthew Arnold: The Critic and France* (Geneva: Librairie Droz, 1964), and S. M. B. Coulling, "Renan's Influence on Arnold's Literary and Social Criticism," *Florida State University Studies, No. 5: English and American Literature* (Tallahassee: Florida State U Research Council, 1952), pp. 95–112. For a discussion of Arnold's understanding and use of science generally, see Fred A. Dudley, "Matthew Arnold and Science," *PMLA,* 57 (1942), 275–94. Although Coulling minimizes the influence of Renan's writings on Arnold, and Dudley concludes that Arnold only incompletely integrated culture and science, both critics recognize that Arnold relied upon ethnology as a science to support his ideas in *On the Study of Celtic Literature* and *Culture and Anarchy.* See Coulling, pp. 102ff., and Dudley, pp. 277, 289.

147. Matthew Arnold, *Culture and Anarchy with Friendship's Garland and Some Literary Essays,* comprising volume 5 of *The Complete Prose Works of Matthew Arnold,* ed. R. H. Super (Ann Arbor: U of Michigan P, 1965), p. 173; hereafter referred to in the text as C & A.

148. Faverty, *Matthew Arnold,* p. 2.

149. Matthew Arnold, *On the Study of Celtic Literature,* in *Lectures and Essays in Criticism,* comprising volume 3 of *Complete Prose Works,* p. 353; hereafter referred to in the text as OSCL.

150. Lionel Trilling points out that Arnold's "own use of the theory contradicts the theory itself. For, having defined the constant qualities of Celtic Blood, and having pointed out that the present English are an admixture which includes the Celt, he proceeds to urge his countrymen to give a greater prominence to their Celtic part." See Trilling, *Matthew Arnold,* p. 236.

151. Arnold's conviction that Hellenism is a scientifically demonstrable force is reflected in his recurrent references to it as a "law" in the concluding portions of each of the last three chapters of *Culture and Anarchy.* In "Porro Unum Est Necessarium" Arnold wishes "to come as near as we can to *the firm intelligible law of things*" (C & A, 191). Two pages from the end of "Our Liberal Practitioners" Arnold says, "Hellenism, surely, or the habit of fixing our mind upon *the intelligible law of things,* is most salutary" (C & A, 219). Finally, in the "Conclusion" to *Culture and Anarchy* there occurs Arnold's most telling use of the phrase: "But we are sure that the endeavor to reach through culture, *the firm intelligible law of things,* we are sure that the detaching ourselves from our stock notions and habits, that a more free play of consciousness, an increased desire for sweet-

ness and light, and all the bent which we call Hellenising, is the master-impulse even now of the life of our nation and of humanity,—somewhat obscurely perhaps for this actual moment, but decisively and certainly for the immediate future; and that those who work for this are the sovereign educators'' (C & A, 229; italics added).

152. He believed that Ireland should be ''a nation poetically only, not politically'' (''From Easter to August,'' *The Nineteenth Century,* 22 [September 1887], 321).

153. Matthew Arnold, ''Renan's 'La Réforme intellectuelle de la France,' '' *God and the Bible,* comprising volume 7 of *Complete Prose Works,* p. 45. Faverty (*Matthew Arnold the Ethnologist,* p. 167) suggests Renan's influence was second only to that of Thomas Arnold. Arnold's political antipathy to the Celts and his disagreement with Renan's theory that the Celts were the political as well as cultural leaders is discussed at length in Harding's *Matthew Arnold: The Critic and France,* pp. 116–19.

154. *The Fortnightly Review,* 55 (n.s. 49) (1 February 1891), 267–77; hereafter referred to in the text as CEA.

155. *Letters,* pp. 286–87. A letter from Allen to Wilde, dated 6 February 1891, shows Allen to be a kindred socialist: ''Will you allow me to thank you most heartily for your noble and beautiful essay in this month's *Fortnightly*? I would have written every line of it myself—if only I had known how'' (*Letters,* p. 287n).

156. *The Woman's World,* 2 (May 1889), 390. In a letter (c. 30 June 1889) to Mrs. R. B. Cunninghame Graham, Wilde commented, ''Seriously speaking, however, what I want to see is a reconciliation of Socialism with Science. Ritchie, in his *Darwinism and Politics,* has tried to do this, but his book, which I suppose you have seen, is very slight and amateurish.'' See Hart-Davis, *More Letters,* p. 84.

157. The pervasiveness of the view is argued in George Gaylord Simpson, *The Meaning of Evolution: A Study of the History of Life and of Its Significance for Man* (New Haven: Yale U P, 1949), pp. 298–99.

158. Important works of the radical Darwinists included: Patrick Geddes and J. A. Thomson, *The Evolution of Sex* (London: W. Scott, 1889); Alfred R. Wallace, *Darwinism: An Exposition of the Theory of Natural Selection with Some of Its Applications,* 3rd edn. (London: Macmillan, 1889), esp. the last chapter; Peter Kropotkin, *Mutual Aid, a Factor of Evolution,* ed. Paul Avrich (New York: New York U P, 1972).

159. George Woodcock's *The Paradox of Oscar Wilde* (London: T. V. Boardman, 1949) has provided a standard, sympathetic, and accurate (as far as it goes) interpretation of Wilde's political position as a libertarian or anarchistic socialist. But Woodcock's analysis must be expanded and modified in the light of Wilde's synthesis of philosophical idealism and science. Woodcock says that ''the reading of Chuang Tsu's writings had a decisive influence on Wilde's own philosophy, confirming his natural tendencies towards non-action and philosophic anarchism'' (p. 148). See also: Edouard Roditi, *Oscar Wilde* (Norfolk, Conn.: New Directions, 1947), p. 150; M. D'Amico, ''Oscar Wilde between 'Socialism' and Aestheticism,'' *English Miscellany,* 18 (1967), 118; and, Isobel Murray, ''Oscar Wilde's Absorption of 'Influences': The Case History of Chuang Tzu,'' *The Durham University Journal,* n.s. 33 (December 1971), 1–13. None of these critics recognizes the ethical implications of Wilde's conception of contemplation or the importance of evolutionary theory.

160. ''A Chinese Sage,'' *The Artist as Critic,* p. 226; hereafter referred to in the text as CS.

161. By missing the paradoxical use of words like ''inaction'' and ''quietism,'' critics have misunderstood the philosophical meaning and ethical import which Wilde attached to contemplation. See our previous discussion of Wilde's rejection of Paterian aesthetic withdrawal in favor of Aristotelian *theoria* in the section in this volume entitled ''Wilde at Oxford: Ruskin's Idealism Versus Pater's Materialism.''

162. For a review of the commentaries about coherence in Wilde's criticism, see Wendell Harris, "Oscar Wilde and Arthur Symons," *Victorian Prose,* pp. 459–67. Relevant discussions include: Roditi, *Oscar Wilde,* chapter eight, esp. pp. 164–67; D'Amico, "Oscar Wilde between 'Socialism' and Aestheticism," pp. 122–23; Philip Rieff, "The Impossible Culture: Oscar Wilde and the Charisma of the Artist," *Encounter,* 35 (September 1970), 33–44; George Watson, *The English Ideology* (London: Allen Lane, 1973), p. 235; Epifanio San Juan, Jr., *The Art of Oscar Wilde* (Princeton: Princeton U P, 1967), p. 5; Brian Nicholas, "Two Nineteenth-Century Utopias: The Influence of Renan's 'L'Avenir de la Science' on Wilde's 'The Soul of Man Under Socialism,' " *Modern Language Review,* 59 (July 1964), 361–70, esp. p. 369; Alick West, *The Mountain in the Sunlight: Studies in Conflict and Unity* (London: Lawrence and Wishart, 1958), pp. 123–53, esp. p. 147; Shewan, *Oscar Wilde: Art and Egotism,* pp. 105–11.

163. For a survey of the anarchist and socialist influences of Kropotkin, Godwin, Morris, the Fabians, and others on Wilde, see J. D. Thomas, " 'The Soul of Man Under Socialism': An Essay in Context," *Rice University Studies,* 51 (Winter 1965), 83–95.

164. See Kropotkin, *Mutual Aid,* pp. 27–31, 245–51. See also Wallace, *Darwinism,* pp. 466, 474. Kropotkin's first chapter appeared in the same issue of *The Nineteenth Century* as the second half of Wilde's "The Critic as Artist [The True Function and Value of Criticism]," 28 (September 1890). Kropotkin also referred to Wallace's theories; the latter's book first appeared in 1889. Kropotkin praised Wilde's "The Soul of Man Under Socialism" as "that article that O. Wilde wrote on Anarchism—in which there are sentences worth being engraved, like verses from the Koran are engraved in Moslem lands" (quoted in *Robert Ross, Friend of Friends; Letters to Robert Ross, Art Critic and Writer, together with Extracts from His Published Articles,* ed. Margery Ross [London: Jonathan Cape, 1952], p. 113).

165. In "The English Renaissance of Art" Wilde defined the artist as man perfected by approvingly quoting William Morris' comment, "I have tried to make each of my workers an artist, and when I say an artist I mean a man" (ER, 479).

166. Wallace, "Human Selection," *The Fortnightly Review,* 54 (n.s. 48) (1890), 325; hereafter referred to in the text as HS. The story of the eugenics movement in late-nineteenth-century Britain is partially told in Daniel J. Kevles' *In the Name of Eugenics: Genetics and the Uses of Human Heredity* (New York: Knopf, 1985). Kevles concentrates on the contributions of Francis Galton and Karl Pearson; most lesser figures get only passing mention.

167. See Grant Allen, "Plain Words on the Woman Question," *The Fortnightly Review,* 52 (n.s. 46) (1889), 448–58; Allen, "The Girl of the Future," *The Universal Review* (May 1890); Wallace, "Human Selection."

168. Quoted in Hyde, *Oscar Wilde,* p. 115. The first version was, as Hyde remarks, "an ingenious piece of Shakespearean research." Wilde's second version adds historical information and biographical speculation which supports this thesis (Shewan, *Art and Egotism,* 86). We base our interpretation on this second version, which appears in CW, 1150–1201.

169. In Hegel's scheme of historical periodization, the modern and Romantic cultural period begins with the Renaissance, so Wilde's use of the term "Romantic" should not be interpreted to mean only literature written after 1790.

170. Earlier versions of this same theory, for example, appear in "The Critic as Artist," where Gilbert speaks of realizing "the collective life of the race" (CW, 1040–41), and in the notebook passages on tribal self (CB, 119–21).

171. For other remarks on art as expression, see CB, 163, and "The Critic as Artist," CW, 1038.

172. "If it were permissible to personify the unconscious, we might call it a collective human being combining the characteristics of both sexes, transcending youth and age, birth and death, and, from having at his command a human experience of one or two million years, almost immortal. . . . He would have lived countless times over the life of the individual, of the family, tribe and people, and he would possess the living sense of the rhythm of growth, flowering and decay." Carl Gustav Jung, *Modern Man in Search of a Soul,* trans. W. S. Dell and Cary F. Baynes (1933; rpt. New York: Harcourt Brace Jovanovich, 1963), p. 186.

173. "It is enough that our fathers believed. They have exhausted the faith-faculty of the species. Their legacy to us is the scepticism of which they were afraid. Had they put it into words, it might not live within us as thought" (CW, 1040).

174. See "The Critic as Artist," CW, 1027–28, where Wilde discusses criticism as the record of one's soul.

175. With his analytic-synthetic method Wilde characteristically explores the power of material forces in history to shape the development of art and ideas. His awareness of how the material medium shapes or even becomes the message can be seen in an early entry in his College Notebook, where he states that "it is impossible to lay down a railway without creating an intellectual influence" (CN, 11). In his essays Wilde applies this idea to both art and politics. In "The Critic as Artist" he comments that men "rage against Materialism, as they call it, forgetting that there has been no material improvement that has not spiritualised the world, and that there have been few, if any, spiritual awakenings that have not wasted the world's faculties in barren hopes, and fruitless aspirations, and empty or trammelling creeds" (CW, 1023). This insight forms the core of his argument in "The Soul of Man Under Socialism." In *Intentions* his spokesmen suggest that the material medium of the artist is the basis for the spirit's realization in culture. In "The Decay of Lying" Vivian says, "The highest art . . . gains more from a new medium or a fresh material than she does from any enthusiasm for art, or from any lofty passion, or from any great awakening of the human consciousness" (CW, 987). In "The Critic as Artist" Gilbert remarks that frankly decorative art "is, of all our visible arts, the one art that creates in us both mood and temperament. Mere colour, unspoiled by meaning, and unallied with definite form, can speak to the soul in a thousand different ways. The harmony that resides in the delicate proportions of lines and masses becomes mirrored in the mind. The repetitions of pattern give us rest. The marvels of design stir the imagination. In the mere loveliness of the materials employed there are latent elements of culture" (CW, 1051). These elements of culture "latent" in the materials of art are perhaps Nature's "good intentions." In any case, Wilde's dialectical assumptions are so synthetic that he appears to be more a soul—albeit an effete and anarchic one—in search of dialectical materialism than a would-be philosopher-king.

176. Wilde's criticism remains grounded on the idealist conception of Hellenic culture outlined in his notebooks and the "Rise." Furthermore, these letters reveal Wilde's continued use of and reliance upon other ideas he explores and develops in the notebooks: Spencer's conception of evolutionary ethics and the social organism, his equation of the good with the healthy, and his belief in the dialectical nature of perception.

177. Wilde recorded "*φιλοσοφειν μετ' ερωτος*" as one of the "two Greek ideals" of education (CB, 40).

178. This would be rather like adopting the utilitarian theory that it is quite all right for the individual to act so as to maximize his individual pleasure, but then to drop its collective goal of achieving the greatest happiness for the greatest number.

179. Paolucci, *Hegel: on the Arts,* p. 4.

180. Basil draws Dorian costumed as Paris, Adonis, and Narcissus (CW, 94).

181. Wilde expresses the same possibility for multiplying and enriching the personality in ''The Critic as Artist'' (CW, 1048), and it is essential to the narrator's development in ''The Portrait of Mr. W. H.'' In both cases, evil is not the only inheritance.

182. Another example of the inheritance of acquired characteristics imagined (with some humor) as a cultural and family phenomenon appears in Wilde's characterization of James and Sibyl Vane, the illegitimate children of a highly connected aristocrat and an actress. Mrs. Vane passes on her melodramatic style to both her children, even though both dislike her histrionics. Sibyl became an actress of aristocratic beauty and James a Hamletlike youth given to passionate threats and ''mad melodramatic words'' (CW, 65). Sybil says to her brother, ''Oh, don't be so serious, Jim. You are like one of the heroes of those silly melodramas mother used to be so fond of acting in'' (CW, 63). To Dorian ''Sybil Vane . . . seemed to be absurdly melodramatic'' (CW, 76). Wilde attributed James's hatred of Dorian to racial inheritance; he ''hated him through some curious race-instinct for which he could not account, and which for that reason was all the more dominant within him'' (CW, 61).

183. The sensuous, aesthetic perception of truth Hegel describes is also expressed in Wilde's ''Preface'' to *Dorian Gray,* which is usually taken as a manifesto of art for art's sake or as a protomodernist document. Its careful use of oppositional terms (e.g., ugly/beautiful, moral/immoral, realism/romanticism, surface/symbol) is really Hegelian and dialectical. One can rise, the aphorisms suggest, to a Hegelian understanding befitting an educated member of the cultural elite, but only if one sheds utilitarian and didactic aims for art (CW, 17). By setting the ideas of the ''Preface'' in relation to the narrative, Wilde problematizes and enriches (or, some might say, mystifies) both and so creates the same kind of ''art-criticism'' as ''The Truth of Masks'' in its revised form.

184. Edward Said, *The World, the Text, and the Critic* (Cambridge, Mass: Harvard U P, 1983), pp. 19–20. Hereafter referred to as WTC.

An Index of Proper Names and Subjects in the Commentary

An Index of Proper Names and Subjects in Oscar Wilde's Commonplace Book (Keyed to Notebook Page Numbers)

An Index of Proper Names and Subjects in Oscar Wilde's College Notebook (Keyed to Notebook Page Numbers)

Metaphysis.

Like the Sphinx those who do not try to answer her questions are annihilated — and often he who solves her riddle comes to no good end — for he dwells apart from men in the colourless abstractions of a spiritual world, in the vacuity of an infinite which has no content.

The danger of metaphysics is that men are often turning nomina into numina.